Discrete Mathematics for Computer Science

DISCRETE MATHEMATICS FOR COMPUTER SCIENCE

Edited by:

Prof. Pomde N.P

ARCLER
PRESS

www.arclerpress.com

Discrete Mathematics for Computer Science

Prof. Pomde N.P

Arcler Press
224 Shoreacres Road
Burlington, ON L7L 2H2
Canada
www.arclerpress.com
Email: orders@arclereducation.com

ISBN: 978-1-77469-830-3 (Hardcover)

ABOUT THE EDITOR

Prof Pomde N. (1984) is presently working as a Assistant Professor, Dept of Agriculture Engineering and Mathematics, College of Agricultural Naigaon Bz,. He has 13 years of experience in teaching and Extension in this duration he has published 02 practical Manual and 02 popular articles. Prof. Pomde has completed M.sc mathematics from Dr. B.A.M. U. university, Aurangabad in 2011.

TABLE OF CONTENTS

LIST OF FIGURES

LIST OF ABBREVIATIONS

AC	Alternating Currents
CDF	Cumulative Distribution Function
CERR	Characteristic Equation of a Recurrence Relation
cos	Cosine
cot	Cotangent
CRT	Chinese Remainder Theorem
CSC	Cosecant
GCD	Greatest Common Divisor
LCM	Least Common Multiple
ODEs	Ordinary Differential Equations
PMF	Probability Mass Function
sec	Secant
sin	Sine
tan	Tangent
ZFC	Zermelo-Fraenkel Set Theory

ABSTRACT

This volume demonstrates that computer science problems can be understood using mathematical models and techniques. Proofs take a key part in this work since the authors share a principle with mathematicians that proofs are fundamental for real knowledge. Proofs likewise play a major role in computer science; they are commonly used to ensure that hardware and software products will always act accurately, something that no level of testing can achieve. In other disciplines, for example, physics, scientific truth is affirmed through experiment. However, in computer science, it is founded through statistical analysis of sample data.

PREFACE

In mathematics, a proof is the verification of a proposition by a series of logical deductions from a base structure of axioms. For computer scientists, one of the main things to demonstrate is the accuracy of programs and systems, whether they do what they are meant to achieve. Programs are becoming highly buggy, and there is a developing community of specialists and experts seeking ways to improve program accuracy. These efforts have been effective enough on computer processor chips that they are currently being used by leading chip producers to demonstrate chip accuracy and prevent errors. Creating mathematical models to confirm programs and systems is still an active area of study.

A proof is defined in this volume as any series of logical deductions from axioms and earlier proved statements that finish with the relevant proposition. Initially, the freedom to develop a proof may seem overwhelming. Different proofs adhere to one of a few standard templates. Each proof consists of its unique details, obviously, however, these layouts at least offer a structure to fill in. A considerable number of these templates actually fit together; one might provide the scientist a high-level structure while others assist at a higher degree of detail.

The demonstration of how algorithms can be validated without the use of a computer is a key application of the Principle of Mathematical Induction in both forms. The idea of a set underscores the vast majority of modern math and software engineering. To involve sets as the foundation for structures in this volume, the authors first offer an understanding of both the language used to depict sets and the tasks commonly connected with sets. The algorithm of sets is exceptionally accurate. When this language is carefully used, we gain accuracy in computer communications and interpreting answers for issues.

Understanding primary operations on sets and task properties is a model for the approach that is developed to introduce most other unique computer structures in this volume. In improving the knowledge of set procedures, readers will understand proof techniques to examine other separate mathematical subjects, e.g., functions, relations, and graphs. The authors use these proof techniques, for instance, to demonstrate that algorithms are accurate and to define how well a certain algorithm has been selected for a particular task.

The language for describing element collections and the set concept is discussed by the authors. Furthermore, they provide a few proof templates that serve as guides for comprehending and constructing proofs. The volume additionally covers common set operations such as unions, complements, intersections, products, and power sets. A few additional proof templates are provided and derived from proofs. Furthermore, it offers a method for numbering the elements in set collections, where these sets might have certain elements that differ across the set. It likewise addresses key proof techniques known as the Principle of Mathematical Induction. The authors use induction for identification of elements whereby an integer statement is valid.

CHAPTER 1

Mathematical Logic and Proofs

Contents

We will look at fundamental proof strategies and how a foundation in propositional logic can help us understand them in this set of notes. Using simple examples, we will show how to apply these proof strategies and demonstrate their effectiveness by utilizing truth tables and other logical devices.

Note: Basic arithmetic properties will be used to illustrate proof ideas throughout these notes. For the time being, assume that mathematics works as you believe it does; we will develop a set of axioms and structure later (Zilber, 2010).

1.1. PROVING CONDITIONAL STATEMENTS

Although proving conditional statements has its own section, almost every proof you will ever write is really just proving a conditional statement. In general, we want to establish the veracity of a $p \Rightarrow q$ statement. Let's look at a simple example to see how we can interpret mathematical statements in this manner.

Example 1: Consider the following statement.

Let a and b be integers. If a is even and a divides b, then b is also even.

We'd like to consider writing this as $p \Rightarrow q$, a single conditional statement. Remember that this can be read as "anytime q must also be true, p must also be true." As a result, we could assign the following values to the propositional variables:

p: (a and b are integers) ∧ (a is even) ∧ (divides b) q: b is even.

The claim we want to prove can then be deduced from these propositional variable assignments as $p \Rightarrow q$.

The simplest way to demonstrate a claim like the one in Example 1 is as follows: Assume proposition p is correct, then demonstrate through a series of logical steps that proposition q must also be correct. The following theorem serves as the structure's foundation:

Theorem 1: $[(p \Rightarrow r) \wedge (r \Rightarrow q)] \Rightarrow [p \Rightarrow q]$

Proof: So as prove this theorem, we have to show that the above proposition is always true. Remember that the conditional statement $p \Rightarrow q$ can also be written as $\neg p \vee q$). Thus, one can rewrite the entire structure as shown:

$$[(p \Rightarrow r) \wedge (r \Rightarrow q)] \Rightarrow [p \Rightarrow q] = [(\neg p \vee r) \wedge (\neg r \vee q)] \Rightarrow (\neg p \vee q)$$
$$= [(\neg p \vee r) \wedge (\neg r \vee q)] \vee (\neg p \vee q)$$

$= [\neg(\neg p \vee r) \vee \neg(\neg r \vee q)] \vee (\neg p \vee q)$ (by DeMorgan's Laws)

$= [(p \wedge \neg r) \vee (r \wedge \neg q)] \vee (\neg p \vee q).$

Hence, in order to prove the theorem true, it suffices to show that $(P \wedge \neg r) \vee (r \wedge \neg q)] \vee (\neg$ p tautology. We consider a truth table (Figure 1.1):

(1)

P	¬P
F	T
T	F

(2)

P	Q	&
F	F	F
F	T	F
T	F	F
T	T	T

(3)

P	Q	v
F	F	F
F	T	T
T	F	T
T	T	T

(4)

P	Q	⇒
F	F	T
F	T	T
T	F	F
T	T	T

(5)

1	2	3	4	5	6	7	8	9	10	11	12	13
P	Q	R	Q	⇒	R	(P	&	Q)	⇒	(P	&	R)
F	F	F	F	T	F	F	F	F	T	F	F	F
F	F	T	F	T	T	F	F	F	T	F	F	T
F	T	F	T	F	F	F	F	T	T	F	F	F
F	T	T	T	T	T	F	F	T	T	F	F	T
T	F	F	F	T	F	T	F	F	T	T	F	F
T	F	T	F	T	T	T	F	F	T	T	T	T
T	T	F	T	F	F	T	T	T	F	T	F	F
T	T	T	T	T	T	T	T	T	T	T	T	T

Figure 1.1. (1–4) relations: ¬, &, v, ⇒; and (5) An inference over formulas: Q ⇒ R ⊢ (P & Q) ⇒ (P & R).

Source: Cheng (2020); Scientific Figure on ResearchGate.

Thus, Tautology and the statement of the theorem are logically equivalent and thus it is itself a tautology. Hence the theorem is true.

This can look like a small thing to prove, but it is important the crux of all mathematical proof. The idea is that if you wish to show that $p \Rightarrow q$ is true, it can be shown by taking a series of implications, having the form.

$p \Rightarrow r1, \quad r1 \Rightarrow r2, \quad r2 \Rightarrow r3, rk{-}1 \Rightarrow rk, \quad rk \Rightarrow q.$

The previous theorem shows that this is enough to prove the statement p ⇒ *q*. Generally, we hope to take these intermediary propositions to be clearly true, or previously proven to be true (Zilber, 2010).

Thus, the basic direct proof structure looks as follows:

Direct Proof of p ⇒ *q*

Assume *p* to be true.

Conclude that *r*1 must be true (for some *r*1).

Conclude that *r*2 must be true (for some *r*2).

Conclude that *rk* must be true (for some *rk*).

Conclude that *q* must be true.

As it will be seen later, propositional logic is not typically used to frame a mathematical proof. However, the structure of propositional logic allows us to know whether the method for proving a statement described above will be effective. Returning to Example 1, consider what this structure might look like. To prove the proposition in this example, we will first write a proof in the above format, then reformat it to fit a traditional proof style (Zilber, 2010).

Example 1-Continued: Recall the statement we wish to prove:

Let a and b be integers. If a is even and a divides b, then b is also even.

We can approach this proof using the structure described above by assuming is true and then drawing a series of inferences that lead us to the conclusion that q is also true (Andrews, 2013).

Assuming p is true, so as a and b are integers, a is even, and a divides b.

By definition, there is an integer k with $a = 2k$, and there is an integer l with $b = al$.

By substitution, we can write $b = al = (2k)l = 2(kl)$.

Since $b = 2(kl)$, b is even

The statements in steps 2 and 3 of the preceding examples can be seen as r1 and r2, and we can see that each of these implications is categorically true by definition or by virtue of fundamental multiplication properties. As we can see, this is structurally consistent with the fundamental concept outlined in our Direct Proof method, as $p \Rightarrow r1$, $r1 \Rightarrow r2$, and $r2 \Rightarrow q$. When they are combined, they demonstrate the entire assertion (Zilber, 2010).

In terms of content, the proof provided here is excellent. However, the proof deviates from accepted mathematical conventions in that it omits the enumeration and presents the proof in a single paragraph; Assume p is true, so that a and b are integers, a is even, and a divides b. By definition, there exists an integer k with $a = 2k$, and there is an integer l with $b = al$. By substitution, we can write $b = al = (2k)l = 2(kl)$. Since $b = 2(kl)$, b is even.

Proofs should include verbal, nounal, and grammatical sentences.

Never start a sentence with a mathematical symbol. Alternatively, always begin a sentence with a word. Because "" can also be a mathematical symbol, you don't want people to think you're performing multiplication when you're simply ending one sentence and beginning another.

It is customary to provide a rationale for any conclusions reached. Things like "by definition," "by arithmetic rules," and so on have already been mentioned. This can help to shed light on the proof's intermediate conclusions. If you can't provide a similar justification for why something is true, it might not be a fair conclusion to reach.

If you'd like to introduce a new symbol, you should clearly define what kind of thing it is. For example, in the proofs in Examples 1 and 2, we introduced variables and specified that these variables represented integers (Zilber, 2010).

1.2. PROVING BICONDITIONAL STATEMENTS

A biconditional statement is a statement of the form $p \Leftrightarrow q$. As seen, $p \Leftrightarrow q$ is logically equivalent to $(p \Rightarrow q) \wedge (q \Rightarrow p)$. Thus, one can approach a proof like this of proposition effectively as two proofs: prove that $p \Rightarrow q$ is true and prove that $q \Rightarrow p$ is true.

It is very common in proofs of biconditional statements to mark the two proofs using the symbols ($\Rightarrow$) and ($\Leftarrow$), to indicate $p \Rightarrow q$ and $p \Leftarrow q$, respectively. It is also common to refer to these types of statements as "if and only ifs," a silly but functional no unification of the operator $\Leftrightarrow$. It is also common to refer to the two parts of the proof as "directions," with $p \Rightarrow q$ called the "forward direction" and $p \Leftarrow q$ called the "backward direction."

A tip for establishing assertions in regard to statements from the previous section. A set of assumptions is often expressed in a proof statement before the thesis to be proved, frequently naming variables and words. In the instance of a simple conditional statement, we merged these presumptions with the premise p. In a biconditional assertion, these presumptions apply to both sides of the proof. We first consider a simple example (Zilber, 2010).

Example 3: Prove the following statement.

Let x be a real number. Define $[x|$ to be the smallest integer greater than or equal to x, and define $[x\rfloor$ to be the largest integer less than or equal to x. Then x is an integer if and only if $[x| = [x\rfloor$.

The initial step in this approach is to decide which presumptions will hold true throughout the evidence. Take note of the word "then" at the beginning of the preceding phrase.

Rather than an assumption, this phrase is frequently used to signify a claim that must be supported. Everything written before the word "then" is an assumption that will be confirmed by the proof, while everything written after the word "then" must be backed by evidence. Given that x is a positive integer and that [x| = [x] is true, the statement "if and only if" expresses a biconditional proposition. Before delving into the meat of the two primary components of the evidence, we can do some "pre-processing" of our assumptions (Andrews, 2013).

Proof: Take x, $[x|$, $[x\rfloor$ as defined in the statement of the proposition. Note that, by definition, we must have that $[x\rfloor \leq x \leq [x|$.

($\Rightarrow$) Assume that x is an integer. Then as $x \leq x$, we must have that the smallest integer greater than or equal to x is x itself, so $[x| = x$. Likewise, the largest integer less than or equal to x is also x itself, so $[x\rfloor = x$. Therefore, $[x\rfloor = [x$($\Leftarrow$) Assume that $[x| = [x\rfloor$. Then since $[x\rfloor \leq x \leq [x|$, and $[x| = [x\rfloor$, we must have that the inequalities are all equalities, so $[x\rfloor = x = [x|$. Since $[x\rfloor$ is an integer by definition, and $[x\rfloor = x$, we must have that x is an integer.

It is important to note that the forward and backward directions of the two propositions to be proved above are treated separately as simple conditional statements, and each is submitted to the direct proof approach described in the prior section. When we build more evidence strategies below, any of these techniques can be utilized to support either of these two propositions (Zilber, 2010).

An issue may contain a biconditional statement that is waiting to be noticed. Consider the following, for instance (Andrews, 2013).

Example 4: Find all real solutions x to the equation $x2 - 2x = 0$.

Solution: First, consider that if x is a solution to the equation,

$x2 - 2x = 0 \Rightarrow x\,(x - 2) = 0$

$\Rightarrow \mathrm{x} = 0$ or $x = 2$.

(You may be tempted to stop right here, but this is insufficient. All that has been demonstrated is that solutions must take the form $x = 0$ or $x = 2$, but we need to also verify that these are, in fact, solutions to the given equation. Indeed, what we have proven thus far is a conditional statement: x is a solution $\Rightarrow x = 0$ or $x = 2$, but we need a biconditional statement here.) Moreover, we find that if $x = 0$, then $x2 - 2x = 0 - 0 = 0$, and if $x = 2$, then $x2 - 2x = 4 - 4 = 0$. Hence, we have that x is a real-valued solution to $x2 - 2x = 0$ if and only if $x = 0$ or $x = 2$.

In this example, a seemingly insignificant algebraic problem conceals a biconditional statement. To tackle the problem of locating all real-valued solutions to an equation, we must first identify the solutions and then confirm that they are the only ones that can be found. We haven't proven that they are even solutions by only establishing that a solution takes the form of x = 0 or x = 2, which is only the first half of the proof. In actuality, we have just answered the second part of the question: we have decided that these are the only possible answers, but we have not confirmed whether or not they are solutions.

We provide one more proof using a biconditional to illustrate the points made above, partly because it is a well-known example and partly because it shows the importance of pre-processing the assumptions before focusing on the two directions of the proof (Andrews, 2013).

1.3. PROOF BY CONTRADICTION

We will construct some more complex proof strategies once we have a basic understanding of direct proof techniques for conditional and biconditional propositions. Here, we start with the evidence by contradiction strategy (Zilber, 2010).

1.3.1. Proving Nonconditional Propositions with Contradiction

To establish a thesis p by contradiction, we frequently begin with the assumption that p is false and then use the direct proof technique to reach a logically improbable conclusion. In essence, we show a statement of this type. $\neg p \Rightarrow q$, where q is never true. Since q cannot be true, one also cannot have $\neg p$ is true, since $\neg p \Rightarrow q$.

Thus, if $\neg p$ is false, one must have that p is true, hence completing the proof of proposition p. Let's look at an example to fathom this method more.

Example 7: Prove the following proposition:

There are no integers a, b for which $2a + 4b = 1$.

Proof: Suppose the proposition is false, so that there are integers a, b for which $2a + 4b = 1$. Dividing both sides of this equation by 2, we conclude that $a + 2b = 1$. Since a and b are integers, $a + 2b$ is also an integer. But 1 is not an integer, so this is impossible (Andrews, 2013).

Hence, the proposition can't be false, it must be true.

Example 8: Prove the following proposition: There is no smallest positive rational number.

Proof: Suppose that the proposition is false, and there is a smallest positive rational number have k as the least positive rational number, so there are positive integers a, b such that $k = a$. Consider $l = k = a$. Notice that since a, b are integers, we also have a, 2b are integers, so l is rational. Also, since a, b is positive, we have that l is positive, and that $l < k$. Therefore, l is a smaller positive rational number than k. Since k is assumed to be the smallest positive rational number, we have arrived at a logically impossible conclusion.

Therefore, the proposition cannot be false, and thus must be true.

The statements in Examples 7 and 8 share one feature: they are both negative. That is, in both cases, we wish to prove the absence of something. This hints at the possibility of using contradiction as a proof approach. It can be difficult to tell whether evidence should be denied, but this method is usually used in this scenario. It's also vital to remember that this might be hidden. For example, terms like "irrational" or "irregular" typically imply contradiction as a viable technique of proof because their definitions are entirely negative: something is irrational if it is NOT rational, and so on (Zilber, 2010).

Generally, a proof by contradiction has to follow this basic structure:

Proof of p by Contradiction

Assume p is false.

Follow the method of Direct Proof to conclude that q must be true (for some q that is observably false).

Conclude that p cannot be false.

Conclude that p is therefore true.

Proposition 1: Let n be an integer. If n2 is even, then n is also even.

Proposition 1 will be used in the example 9 proof, but the demonstration of this proposition is left to the reader. Proposition 1 might be proven just by contradiction.

Example 9:

9 $\sqrt{2}$ is irrational.

Proof: Suppose that the proposition is false, so

$\sqrt{2}$ is rational. Then there exist integers a, b so that $2 = a^b$. We assume that a and b are chosen to have no common factors; that is, the rational b^a is in lowest terms.

By squaring both sides, we therefore have that $2 = a^2\ b^2$, so $2b^2 = a^2$. Therefore, a^2 is even, and hence a must also be even. Thus, there exists an integer k so that $a = 2k$, and $a^2 = 4k^2$.

We therefore have that $2b^2 = a^2 = 4k^2$ and dividing by 2 yields $b^2 = 2k^2$. Therefore, b^2 is even, and hence b must also be even. Since a and b are both even, they are both divisible by 2. But by assumption, a and b have no common factors, so this is impossible (Zilber, 2010). As a result, the statement must be true because it cannot be untrue. As a result, illogical.

We wish to show a statement of the form $p \Rightarrow q$ similar to how simple conditional statements are shown. Remember that this assertion is logically equivalent to the previous set of notes' $(\neg p) \lor q$. This can now be rewritten like this:

$$(\neg p) \lor q \equiv \neg\neg((\neg p) \lor q)$$

$$\equiv \neg(\neg((\neg p) \lor q))$$

$$\equiv \neg (p \land (\neg q)) \quad \text{(by De Morgan's Laws)}$$

That is to say, $p \Rightarrow q$ is true if and only if $p \land (\neg q)$ is false. This allows us to recast any conditional assertion as a negative using the proof by contradiction technique explained in the prior section. This is typically performed by assuming that $p \land (\neg q)$ is true and then reaching an unreasonable conclusion. It must be the case that $p \land (\neg q)$ is true, as we desired, because $p \land (\neg q)$ cannot be true.

Thus, if $p \Rightarrow q$ is true, hence every time p is true, q is also true. Proof by contradiction assumes that p is true, but q is false, resulting in an illogical conclusion. Because it is irrational for q to be false, if p is true, q must be true as well (Andrews, 2013).

Example 10: Let n be an integer. If $n2 + 5$ is odd, then n is even.

Proof: Suppose, for the sake of contradiction, that $n2 + 5$ is odd and n is also odd. By definition, then, there exists integers k and l so that $n2 + 5 = 2k + 1$ and $n = 2l + 1$. Hence, we have

$$2k + 1 = n2 + 5$$

$$= (2l + 1)2 + 5$$

$$= 4l2 + 4l + 1 + 5$$

$$= 2(2l2 + 2l + 3).$$

Therefore, $2k + 1$ is even. This is clearly impossible, and hence we cannot have that $n2 + 5$ is odd and n is also odd.

Therefore, if that $n2 + 5$ is odd, we must have n is even.

The following approach is commonly used to prove conditional claims using contradiction:

Proof of $p \Rightarrow q$ by Contradiction

Assume that q is false, and p is true.

Use the Direct Proof approach (for some r that is observably untrue) to come to the conclusion that r must be true. Draw the conclusion that q cannot be untrue if p is true (Zilber, 2010).

Draw the conclusion that p ⇒ q whenever p is true, and q is true as well.

As an alternative way of thinking about this, consider the truth table corresponding to p ⇒ q. The proposition p ⇒ q is a false proposition if p is true and q is false, but a true proposition in all other circumstances, according to the definition in the final set of notes. As a result, the truth table looks like this:

p	q	$p \Rightarrow q$
T	T	T
T	F	F
F	T	T
F	F	T

When employing a proof by contradiction, it is critical to consider the second row of the table. If we can show that giving p and q's truth values to the second row is not possible, the truth values must come from one of the other rows. If the second row is illogically impossible, then $p \Rightarrow q$ must be true in any row where $p \Rightarrow q$ is true (Andrews, 2013).

This section concludes with two more examples of contradiction-based implications proofs. In each of the following cases, there is a clue that a contradiction proof would be useful: Both of the following assertions contain a false conclusion q; that is, we want to demonstrate that nothing happens. Similar to how nonconditional propositions are demonstrated by contradiction, it is frequently useful to show that if something happens, something logically impossible must also happen (Zilber, 2010).

Example 11: Prove the following proposition:

Let a and b be integers. If a ≥ 2, then a does not divide one of b and b + 1.

Note: There is the implication p ⇒ q, where p is "a ≥ 2" and q is "a does not divide one of b and b + 1." To follow proof by contradiction, lets to assume p and ¬q. Note that to negate q, one can use De Morgan's laws; think of q as "(a does not divide b) ∨ (a does not divide b + 1)." De Morgan's laws then say that ¬q can be written as

¬q ≡ ¬(a does not divide b)∧ ¬(a does not divide b + 1).

The negation of "does not divide" is clearly "does divide," so to negate q we assume that a divides both b and b + 1.

Proof: Suppose, for the sake of contradiction, that a ≥ 2 but a does divide both of b and b + 1. Then there are integers k, l such that b = ak and b + 1 = al. By substitution, thus have ak + 1 = al, so 1 = a(l – k). Because a ≥

2, we have that a/= 0, so we can divide by a on both sides, to obtain 1 = l – k. Since l and k are integers, l – k is also an integer, but since a ≥ 2, 1 is not an integer. This is impossible.

Thus, it must be that if a ≥ 2, a must not divide at least one of b and b + 1.

Example 12: Prove the following proposition:

If a, b, c are all odd integers, then there is no rational x such that ax2 + bx + c = 0.

Proof: Suppose, for the sake of contradiction, that a, b, c are all odd integers, but that there is a rational x such that ax2 + bx + c = 0. Let k, l be integers with x = k, and suppose that k and l share no common factors, so that x is written in lowest terms. Then we have

$Ax^2 + bx + c = 0 \Rightarrow akl^2 + bkl + c = 0$

$\Rightarrow$ ak 2 + bkl + cl 2= 0 (by multiplying by l2)

Not the right-hand side of the final equation is 0, which is even, so the left-hand side should also be even. As a, b, c is all odd, the observation is that this can only occur if both k and l are even. Thus, both k and l are divisible by 2. But this is impossible, since we have chosen k and l to share no common factors (Andrews, 2013).

Thus, if *a, b, c* are all odd integers, there cannot exist any rational number *x* such that $ax^2 + bx + c = 0$.

A word of caution regarding the preceding example: if you do not believe the argument that k and l must both be even, then prove it! You can demonstrate that a, b, and c are odd numbers, that k and l are integers, and that ak2 + bkl + cl2 = 0 implies that k and l are even. Contradiction can be used to prove this point. Assume they aren't; in that case, at least one of them is odd, and you can end up with a contradiction.

This part of the proof is omitted to avoid complicating the structure, but it is also a helpful exercise to ensure you understand how to construct a proof by contradiction!

As you may have noticed, the reader is always made aware at the start of our proofs by contradiction that we are expressly supposing the assertion to be false or that we intend to proceed by contradiction.

This is common practice: if you don't explain how you aim to achieve contradiction, the reader may not understand why you made an assumption that appears nonsensical based on the wording of the intended assertion (Zilber, 2010).

1.4. PROOF BY CONTRAPOSITIVE

As with proving conditional statements by contradiction, a proof by contrapositive relies on the fact that:

$p \Rightarrow q$ is logically equivalent to $\neg (p \wedge (\neg q))$ but takes a slightly different approach to the proof. Consider:

$$p \Rightarrow q \equiv \neg(p \wedge (\neg q))$$

$$\equiv \neg((\neg q) \wedge p)$$

$$\equiv \neg((\neg q) \wedge \neg(\neg p))$$

$$\equiv (\neg q) \Rightarrow (\neg p)$$

In other words, if $p \Rightarrow q$, then q must also be true. As a result, if q is false, p cannot be true. This is akin to saying that the honesty of ¬q implies the truthfulness of ¬p. Proof by contrapositive is a method of demonstrating conditional assertions. Specifically, demonstrating q⇒ $\neg p$ suffices to demonstrate $p \Rightarrow q$. Any method can be used to accomplish this, albeit if a contrapositive is needed, a direct proof approach is commonly used.

Example 13: Prove the following proposition:

Let a, b be integers. If ab is even, then at least one of a or b is even.

Proof: We work by contrapositive. Suppose that a and b are both odd. Then there are integers k and l so that a = 2k + 1 and b = 2l + 1. Therefore, we have:

ab = (2k + 1)(2l + 1) = 4kl + 2k + 2l + 1 = 2(2kl + k + l) + 1,

so, ab is odd.

Hence, by contrapositive, if ab is even, we must have at least one of a or b is even (Zilber, 2010).

Proof of p ⇒ q by Contrapositive is generally as follows:

1. Assume q is false.
2. Follow the method of Direct Proof to conclude that p is also false.
3. Conclude that ¬q ⇒ ¬p is true.
4. Since (¬q ⇒ ¬p) ≡ (p ⇒ q), conclude that p ⇒ q is true.

Students frequently confuse a proof by contrapositive and a proof by contradiction since the first premise in both cases plainly asserts that q is false. However, there is an important differential in this scenario. When utilizing a proof by contrapositive, you have a specific goal in mind: assuming q is untrue, you wish to show that p is incorrect. The goal of a proof by contradiction is to arrive at any logically impossible conclusion while assuming that q is false, and p is true. Because there are numerous theoretically impossible outcomes, proofs by contradiction have a less specified goal than demonstrations by contrapositive (Andrews, 2013).

However, why use a proof by contradiction rather than a proof by contrapositive? Why take that route when a hazy aim makes proof appear, well, harder? It's an excellent question, and I would suggest you investigate whether you could simply operate by contrapositive instead of contradiction every time you start a proof. Despite this, contradiction is a helpful method since it involves more initial assumptions than proof by contrapositive. As a result, while constructing conclusions, you have more data at your disposal than in a demonstration by contrapositive (Zilber, 2010).

1.5. PROOF BY CASES

Our initial analysis of proof by cases will focus on explicitly stated cases in a proposition's statement. Consider the proposition $p \Rightarrow q$, which may be expressed as a propositional formula with the disjunction operator because p takes the form $p \Rightarrow s$. To explore a claim like this using the Direct Proof method, we must first assume that p is true. But if p is correct, we only have two choices: either r or s are correct, or both. These possibilities can be subdivided into instances, yielding two proofs. By examining the following truth table, we see that $(r \lor s) \Rightarrow q$ is logically equivalent to $(r \Rightarrow q) \land (s \Rightarrow q)$, so this approach of proving two separate cases is sufficient to prove the proposition.

r	s	q	$r \lor s$	$(r \lor s) \Rightarrow q$	$r \Rightarrow q$	$s \Rightarrow q$	$(r \Rightarrow q) \land (s \Rightarrow q)$
T	T	T	T	T	T	T	
T	F	T	F	F	F	F	
F	T	T	T	T	T	T	
F	F	T	F	F	T	F	
T	T	T	T	T	T	T	
T	F	T	F	T	F	F	
F	T	F	T	T	T	T	
F	F	F	T	T	T	T	

In plain English, it is sufficient to consider each of the two alternatives for p individually if p is a proposition that includes a "or" statement.

A proof by cases is now uncommon in this manner. It is common for the proofreader to have to describe scenarios on their own, often based on some key feature of the objects in question. This typically means employing the Law of Excluded Middle from the previous set of notes. In other words, we can contemplate splitting up in accordance with a statement r, where $r \vee (\neg r)$ is self-evidently true. Consider the following scenarios: An integer can be either even or odd, and a number can be either negative or positive (Zilber, 2010).

Example 15: Prove the following proposition: If x is a real number, then $|x + 3| - x > 2$.

Proof: We consider two cases: $x \geq -3$ and $x < -3$.

Case 1: $x \geq -3$. Then $|x + 3| = x + 3$, so we have $|x + 3| - x = x + 3 - x = 3 > 2$, so the proposition holds.

Case 2: $x < -3$. Then $|x + 3| = -(x + 3)$, so we have $|x + 3| - x = -3 - x - x = -3 - 2x$. Since,

$x < -3$, we must have $-x > 3$, so $-3 - 2x > -3 + 2(3) = 3 > 2$. Therefore, the proposition holds.

Since the proposition holds in all cases, it must be true that if x is a real number, then $|x+3|-x > 2$.

In the above example, there is a very compelling reason to separate the proof into individual scenarios. We know that the absolute value function has cases: if the argument is nonnegative, we take one number; if the argument is negative, we take a second number. As a result, considering proof by cases makes sense (Andrews, 2013).

In some circumstances, though, using only two cases may not be adequate. On some occasions, it may be preferable to divide proposition p into several cases and prove each one separately. This is also appropriate, as seen in the following illustration; $\vee \quad \vee \cdots \vee rk$, so long as we can be sure that $p\ r1 = r2$, then we will have

$$p \Rightarrow q \equiv (r1 \vee r2 \vee \cdots \vee rk) \Rightarrow q \equiv (r1 \Rightarrow q) \wedge (r2 \Rightarrow q) \wedge \cdots \wedge (rk \Rightarrow q),$$

that is, we can prove each $ri \Rightarrow q$ independently.

1.6. VARIABLES AND QUANTIFICATION

The explanation of variables concludes this section of proofreading suggestions notes. Variables have been mentioned several times in these notes, but there hasn't been a full description of how to determine what kinds of values these variables might have. We discussed the need of clearly specifying each variable used in a problem from the start of the first set of notes, and we've seen that throughout these notes: each time a letter appears, it's mentioned what kind of value it can take (an integer, a rational, etc.) (Zilber, 2010).

1.6.1. Units of Quantification

For the most part, we've enabled our variables to accept any value from the set they came from, so they might be any integer, any odd integer, and so on. Let's make this more formal.

Given a variable x, the range of x is the set of possible values that x can take. If the range is X, we write x ∈X to indicate that x is a member of X. If x is permitted to take any value in its range, then we say that x is universally quantified.

In most of our examples to this point, then, we have used universal quantification. That is, we have phrased our propositions in the form "Let x be in range X. Then [proposition about x]." Here are several other common rephrasings of this proposition:

Let x ∈ X. Then [proposition about x].

For all x ∈ X, [proposition about x]. Given x ∈ X, [proposition about x]. All of these constructions indicate that the proposition we wish to prove is universally true, that is, it applies to every member of the range, no matter what it looks like. In all of these cases, our proofs can use no information about x beyond the fact that it is a member of the specified range (Andrews, 2013).

Symbolically, we express universal quantification with the symbol ∀. This symbol is read as "for all" or "for any." So, we could express the statement.

Let a, b integers. Then a + b is also an integer.

Using the symbols ∀a, b ∈ Z, a + b ∈ Z. (Here, Z is a symbol representing the set of integers)

Now that we have a clearer understanding of universal quantification, we may recast many things that we previously thought of as conditional assertions in the context of propositional logic as nonconditional, universally quantified propositions (Zilber, 2010).

For example, consider the proposition.

For all integers a, $a2 + a$ is even. Up until now, we may have viewed this as conditional, under the structure p: "a is an integer," q: "$a2 + a$ is even," we can see this statement as $p \Rightarrow q$. However, we could redefine this structure by taking a logical proposition that has a variable; that is, let p(a) be the logical statement $a2 + a$ is even. We can then rephrase the structure of this proposition as

$\forall a \in Z, p(a)$.

In this way we can see many of the propositions we have handled thus far as in fact propositional formulae containing variables, as above. Formally, we have the following:

Definition 2: Let p(x) be a logical formula that takes a variable x from range X. The proposition "$\forall x \in X, p(x)$" is true if p(x) is true for every choice of $x \in X$, and false otherwise (Andrews, 2013). It is worth mentioning here that this gives us an inkling of how to disprove statements of the type $x\forall \in X$, p(x). Since this proposition can only be true if it is in fact true for every choice of x, then all that is needed to prove this proposition false is to demonstrate ONE value of x for which the proposition p(x) fails to hold. For example:

Example 16: Disprove the following proposition: For all real numbers x, $x2 > 0$.

Proof: Let $x = 0$. Then x is a real number, but $x2 = 0$, which is not greater than 0. Therefore, the proposition is false. It makes no difference in the above example whether $x2 > 0$ holds true for any of the other potential values of x. It simply needs to be false once to show that it does not hold "for ALL real numbers." It is sufficient to demonstrate its untruth by referring to a single instance of x outside of the allowed range (Zilber, 2010). Just a reminder: do not use the term "random" to describe a variable that has been statistically quantified across the board. Although the phrase "for any random integer" is commonly used, it has a deeper, subtler meaning in mathematics. Instead of using the symbol, you could be more descriptive about your quantification by using the word "arbitrary" rather than "random," which is both valid and relieves the reader of the burden of worrying about probability distributions (Andrews, 2013).

1.6.2. Existential Quantification

Although as seen relatively little existential quantification thus far, it is equally important to understand mathematical structure. First consider what a proposition that has an existentially quantified variable looks like.

Definition 3: Let $p(x)$ be a logical formula that takes a variable x from range X. The proposition

"$\exists x \in X$, $p(x)$ is true if $p(x)$ is true for some value of $x \in X$, and false otherwise. We read the symbol $\exists$as "there exists" or "for some." Compared to universal quantification, a proposition of this type only requires that $p(x)$ is true for at least one choice of x. To prove a proposition of the form $\exists x \in X$, $p(x)$, it suffices simply to demonstrate one value of x for which the statement $p(x)$ is true. Consider the following example.

Example 18: Prove the following proposition:

There exists a real number x such that x(x + sin x – x^2 + sin2 x cos x + ex) = 0.

Proof: Let x = 0. Then x(x+sin x–x +2 sin x^2 cos x+e)x= 0(0+sin 0–0 +2sin 02 cos 0+e)0 = 0. Hence there exists an x that satisfies the proposition.

In this example, it would be awful to have to spend time doing algebra on x(x+sin–x x^2 +$\sin^2$ x cos x+ e^x) for an arbitrary x. Fortunately, though, we don't have to: because this is an existentially quantified x, it is enough to just demonstrate a single value of x that makes the proposition true. This is easy to do, since 0 times anything is 0 (Zilber, 2010).

1.6.3. Universal and Existential Quantifiers Are Friends

The universal and existential quantifiers play quite nicely together and help each other out. We have already seen one small example of this: to disprove a statement $\forall$ x∈X, p(x), it suffices to show just one x with ¬p(x). This is, effectively, an existential question, and it is formalized with a new version of De Morgan's Laws just for quantifiers (Andrews, 2013).

Theorem 2 (De Morgan's Laws for Quantifiers): Let p(x) be a logical formula that takes a variable x from range X. Then:

1. $\neg(\forall x \in X, p(x)) \equiv \exists x \in X, \neg p(x)$.
2. $\neg(\exists x \in X, p(x)) \equiv \forall x \in X, \neg p(x)$.

The proof of this theorem is left as an exercise and is omitted. The proof is mostly definitional: if p(x) does not hold for all values of x, then there must be some x choice for which p(x) is false (part 1). Section 2 is similar to the previous assertion in that p(x) must always be false if there is no x for which it is true. In Example 16, we applied De Morgan's Laws for Quantifiers subtly rather than openly by employing logical reasoning rather than the theorem.

In addition to supporting one another in negations, the universal and existential quantifiers frequently occur together in a single claim. You may recall this from calculus (or you may have shut it out; it doesn't matter) (Zilber, 2010).

We shall not dwell on this) a definition of continuity that took the form "$\forall\varepsilon > 0$,

$\exists\delta > 0$ such that"

Here, and in many other kinds of circumstance, a formal statement of the proposition we wish to consider requires two variables to be quantified, in different ways. And there is a critical point to be made here: ORDER MATTERS. When we read quantifiers in a mathematical statement, we always give precedence to the first quantifier. If you make a statement of the form "$\forall x \in X$, [more words next]," it is assumed that whatever words follow. $\forall x \in X$ should actually be true for all x in the range of X. Hence, there is a big difference between a statement that takes the form "$\forall\varepsilon > 0$, $\exists\delta > 0$" and one that takes the form "$\exists\delta > 0$, $\forall\varepsilon > 0$." As an illustration, consider a case that has nothing to do with calculus. Let's look at two statements using existential and universal quantifiers and see how they change when the quantifiers are employed in a different order (Andrews, 2013).

1. $\forall a \in Z$, $\exists b \in Z$, $a + b = 0$.
2. $\exists b \in Z$, $\forall a \in Z$, $a + b = 0$.

In answer to the first claim, we state that: First, choose an arbitrary number a. Then, based on a, select a certain integer b. Because the existential quantifier follows the universal quantifier (in this case, $b = -a$), we can use a to pick b. This makes sense in this context and is correct: for every number a, there is surely an integer b that meets the criteria that $a + b = 0$. The second statement, however, is where things begin to go wrong. Because the existential quantifier comes first, we must choose b first. That is, before we can evaluate a, we must first decide on b. Then, once we've decided on b, we want the statement "$a + b = 0$" to be true for EVERY integer a. This is obviously illogical because $a + b = 0$ can only apply to one value of a and not to all potential values of a (Zilber, 2010).

This acts as a warning and a reminder to consider precedence order when thinking about quantified variables. It's also worth noting that this type of problem only occurs when the order of a universal and existential quantifier is reversed; otherwise, we can freely swap two variables with existential or universal quantification. Only when we have only one of each may there be problem.

CHAPTER 2

Basic Mathematics on the Real Numbers

Contents

2.1. INTRODUCTION

Real numbers, symbolized by "R," contain rational and irrational numbers. Rational numbers are fractions of two integers (p/q) with non-zero q. "Q" represents rational numbers (Skudrzyk, 2012)

- Integers (–3, 0, 5);
- Fractions (1/2, –3/4, 7/8);
- Repeating decimals (0.333…, 0.666…);
- Terminating decimals (0.25, 1.75).

Irrational numbers cannot be fractions and have non-repeating, non-terminating decimal expansions. They are not integer fractions. Irrational numbers are indicated by "I."

- √2 (1.41421356…);
- π (about 3.14159265);
- e (Euler's number, about 2.71828183…).

2.1.1. Real Number Characteristics

- Real numbers are positive, negative, or zero.
- The number line shows positive numbers to the right of zero, negative numbers to the left, and zero in the middle.
- Ordered real numbers can be compared using < and > symbols.
- Real numbers are continuous and infinite, hence there are infinite real numbers between any two different real numbers.

2.1.2. Real Number Properties

- Real numbers have important mathematical features for operations and equations.
- Closure Property: Real numbers can be added or multiplied.
- Associative Property: Number grouping does not influence addition or multiplication. $(a + b) + c$ equals $a + (b+c)$.
- Commutative Property: Number order does not impact addition or multiplication. $a+b$ equals $b+a$.
- Multiplication over adds. $a \times (b + c)$ equals $a \times b + c$.

2.2. REAL NUMBER NOTIONS AND OPERATIONS

2.2.1. Addition (+)

2 + 3 equals 5. Math adds two or more numbers. Sum is the outcome of adding addends. We use it daily. Addition occurs frequently. Bills and receipts are prominent instances of adding numbers. The addition has endless daily uses. Cooking, calculating shopping expenditures, calculating distances, and more involve addition. "+" is the addition symbol. 5+3 is "5 plus 3." Addition sentences display the total of two or more values. Addends are the additional values. Addition sentences have two or more addends, plus symbol(s), equal sign, and sum (Skudrzyk, 2012).

2.2.1.1. Solve Addition Problem

The methods below make adding two numbers easy. These approaches simplify adding tiny numbers.

1. **Fingers:** These may add little numbers.
2. **Number Line:** Add with the number line.
3. **Number Grid:** Number chart adds numerals too.
4. **Counting Forward:** This method works in another case. Add 5 and 4.

- **Step 1:** Think big. Here, 5.
- **Step 2:** Repeat the second number four times.

This: 6→7→8→9.

Get 9. Thus, 5 + 4 = 9.

1. **Vertical Addition (Column Method):** Add numbers vertically to solve the issue. We arrange the numbers vertically by place value—ones, tens, hundreds, thousands, etc. Add from the right (digit at one or unit place). Regrouping and non-regrouping may occur when addressing such issues.

2.2.1.2. Addition Properties

1. **Addition Commutative:** Swapping numbers in an addition equation doesn't change the sum (Skudrzyk, 2012).

   19+14=33

   14+19=33

2. **Addition Identity:** Zero and a number add up to the number.

   15+0=15

   0+15=15

3. **Associative Addition:** The sum of three or more numbers is independent of their order. We might add 5, 8, and 6 in many ways:

   (5+8) +6=13+6=19

   5+(8+6) =5+14=19

   8+(5+6) =8+11=19

4. **Distributive Property:** The sum of two numbers multiplied by a third number equals the total of the products received when each number is multiplied by the third number.

   Example: (2+3)5.

   (2+3)5=55=25

   (2+3)5= (25) + (35) =10+15=25

- **Fun Fact:**
  - Zero adds the number;
  - A number's successor is 1 added;
  - The total is unaffected by addend order.

2.2.2. Subtraction (–)

Example: 7 – 4 = 3, If we buy ice cream for $140 and give the cashier $200, we may subtract one real number from another. Subtracting 200–140=60, the cashier returns the surplus. The cashier returns $60. Subtraction is used in everyday life (Skudrzyk, 2012).

If we have 3 candies and our friend asks for 1, how many do we have left? Just remove 3 from 6. Subtraction is the difference between two numbers or quantities. Taking a number away from another is called subtracting. Paying bills, sending money to pals, and more utilize subtraction. Mathematically, operators have different symbols. +, –, /, *, and more. Subtraction is one of the most essential arithmetic symbols. We subtracted 6 and 3 above. Subtraction is seen in this expression: (6–3=3). The minus sign is this symbol (Saleri, 2010).

1. **Subtraction Formula:** We utilize subtraction words when subtracting two numbers:

- **Minuend:** The removed number is the minuend.
- **Subtrahend:** The number deducted from the minuend.
- **Difference:** The subtrahend-minuend difference is the ultimate outcome.

2. **Uses of Minus Sign:**

- **Subtract:** Minus signifies subtraction. Minus sign denotes subtract, decreased by, take away, and how much one value is greater than another.
- **Negative Integers:** Positive, negative, and 0 numbers are integers. Negative integers (no fractions) are represented by the minus sign. A whole number with a negative sign represents a negative integer. Negative integer 5 is (–5).
- **Measuring:** Temperature measurements utilize the negative sign. –4°C is 4° below zero. Another example: 5°C becomes –10°C. Current temperature?
- **Represent Opposite Directions:** The minus symbol represents a negative direction on graph paper. The graph slopes downward.
  - First-quadrant coordinates are (x, y).
  - Second quadrant coordinates are (–x, y).
  - Third quadrant coordinates are (–x, –y).
  - Fourth quadrant coordinates are (x, –y).

3. **Integer Operations with "Minus" Sign**

Negative numbers multiply to positive numbers (Skudrzyk, 2012).

Negative × Negative= Positive

Example: (–5) x (–15) =+75.

Negative and positive numbers multiply to a negative number.

Negative × Positive equals Negative

Example: (–5) x (15) =–75.

Negative numbers always add to negative numbers.

Example: (–3) + (–4) = (–7)

Subtracting a positive number from a negative number always yields negative.

We count backwards from a negative number while subtracting a positive number.

Negative-Positive = Negative

Subtracting two negative numbers. A double negative sign becomes a positive sign. Instead of subtracting, you add a positive. Number magnitude determines the answer (Saleri, 2010).

Negative + Positive

Add the values after transforming – (–5) into +5.

Example: (–2) – (–5). It reads "negative two minus negative 5." (–2) + 5 becomes (–2) +5.

–2 on the number line.

- Negative numbers minus positive numbers always provide positive numbers.

Subtracting a negative number from a positive number of yields a plus sign. Instead of subtracting, you add a positive. The equation becomes an addition issue.

Positive + Positive

The –(–4) becomes +4. Number line starts at 2.

4. **Subtraction Methods:** Subtraction techniques vary.

- **Imagery:** Diagrams may depict what you start with, what you take away, and what you're left with.
- **Number Line Subtraction:** 5 minus 2 starts with 5. Subtracting 2 requires two steps back. We conclude at 3.

Column Method: The column technique of subtraction divides numbers into ones, tens, hundreds, etc., and writes the minuend above the subtrahend. This approach subtracts ones from right to left (Skudrzyk, 2012).

Math Regrouping: Math regrouping is making/breaking groups when adding and subtracting. Regrouping groups in situ value performs an operation. When the minuend digits are smaller than the subtrahend digits, we regroup in subtraction. Regrouping involves reordering integers by place value. Regrouping in subtraction is termed borrowing (Saleri, 2010).

Regrouping Subtraction: Subtraction occasionally involves regrouping. To subtract using the column approach when the bottom digit is bigger than the top digit, we regroup. This regrouping example helps us grasp subtraction by solving 31–19.

We remove the bottom slot's unit place digit from the higher slot. Regrouping—borrowing—occurs when the lower slot number is greater than the higher slot number. In this case, we subtract one from the tens place digit from the upper slot number and write the remaining number above it, that is, we take 1 from 3 and write 2 above 3, while this 1 is "borrowed" to the unit place and made a 10 and added to the unit place number, giving us a two-digit number. 10 is added to the unit place digit from the tens place digit. 10 is added to the unit place digit, 1, and 11 is written above it. We now subtract the two numbers. Subtracting 11–9 from the unit place number of the bottom slot gives us 2. We usually take the leftover number from the higher slot, 2, and remove the lower slot number, 2–1, to get 1, which gives us 12 (Skudrzyk, 2012).

5. **Subtraction Properties:** Subtraction has several key aspects in daily life.

Commutative Subtraction: Swapping numbers does not change the result. Subtraction works differently when minuend replaces subtrahend and vice versa. Subtraction cannot be commutative.

8–5 vs. 5–8.

Subtraction Identity: Identity property indicates that subtracting "0" from an integer yields the number.

Example: 5–0=5.

Subtracting an integer from itself: Subtracting an integer from itself always yields "0."

A–A=0

9–9= 0.

Equal Subtraction: The property states that an equation remains equal if we remove any integer from both sides (Saleri, 2010).

Algebra equation;

$\Rightarrow \times - 3 = 5$

The equation holds if we remove the same amount on both sides. Subtract 8 from both sides.

$\Rightarrow \times - 3 - 8 = 5 - 8$

$\Rightarrow \times - 11 = -3$

Distribution of Subtraction: The property states that multiplication of subtraction equals subtraction of individual number multiplication.

A×(B–C) =A×B–A×C

3x (5–2) =3x3=9 and

2.2.3. Multiplication (×)

5 × 6 equals 30. Multiplication, along with addition, subtraction, and division, is arithmetic. Multiplication involves adding equal-sized groupings. Multiplication equations can be read several ways.

Example: 2 × 3 = 6. It reads:

- Two times three equals six.
- Two three times equals six.

Crosses, asterisks, and dots indicate multiplication. Most notebook writers utilize the cross. Algebra and computer languages employ asterisks and dots (Skudrzyk, 2012).

Example: 6 × 5 = 30

7∗8=56

1. **Multiplying Integers:** To multiply numbers, we require their signs.

Multiply Positive Integers: Positive integers always multiply to positive integers.

Example: 5 × 6 = 30

Multiply a Positive and Negative Integer: A positive and negative integer product is always negative.

(– 5) x6= (– 30)

Multiply Negative Integers: Negative numbers always multiply to positive integers.

(– 5) x (– 6) =30.

Multiplying Fractions: Multiplying fractions involves multiplying numerators and denominators.

a/b x c/d=axC/BxD

Multiplying Decimals: Multiplying decimals equals multiplying integers.

Multiply 13.2 and 3.5.

Multiply 13.2 and 3.5 as whole numbers without decimals. Thus, 132×35 = 4620 (Skudrzyk, 2012).

If we put the decimal point back, the product of two decimal values will contain decimal up to two places from right to left, such that 13.2×3.5=46.20

Power Multiplication: Powers are base times exponent in arithmetic 3^2, $(-5)^4$.

When Power is Different But Base is Identical: x^n and x^m are the expressions. "X" is the base. Multiplying words with the same base adds powers, 32 $3^5 = 3^2 + 5 = 3^7$

The Base is Different But the Exponent is the Same: Let x^m and y^m be m-powered expressions. The bases are multiplied first, then the exponent, when multiplying terms with the same exponents. $3^2 4^2 = (3\times4)^2 = 12^2 = 144$

Different Bases and Powers: X^n and y^m. X and y are bases. Powers are n and m. Each expression is tested before multiplying.

$X^n Y^m = (x)^n (y)^m$ $3^2 4^3 = 9\times64 = 576$

2. **Multiplication Properties:** Multiplication follows the same characteristics as addition:

Commutative Property: Multiplying two integers does not affect the product.

Example: 2×3=6. Calculating 3×2 in reverse yields 6 as well.

Associative Property: Multiplying three or more numbers in order does not matter. If we have 2,3, and 4:

- (2×3) ×4=24
- 2× (3×4) = 24

Changing the order and multiplying won't affect the result (Saleri, 2010).

- 3× (2×4) =24

Distributive Property: Multiplying a number by the sum of two numbers yields the same result as multiplying it by those two numbers separately. Like 3×8 (Skudrzyk, 2012).

- 8=6+2. Thus 3×8=3× (6+2) =24.
- Now, 3×6=18. Also, 3×2=6.
- 18+6=24=3×8.

Using Properties: If you understand multiplication properties, you can break down difficult problems into simpler ones. Such as: 3 × 13 = 3 × (10 + 3) = (3 × 10) + (3 × 3).

This helps create new facts from established facts.

Such as:

- The commutative property of multiplication shows that 9×2 = 18 if 2×9 is 18.
- The distributive principle of multiplication lets you calculate 2×14 from 2×10 and 2×4.

2.2.4. Division (÷)

Division is the inverse of multiplication. After dividing the original 12 into three groups, each group has four people. When splitting anything evenly, the number of groups is determined by dividing it. Place four doughnuts in each of the three equal piles. 12/3 = 4. Indicators of division include:

- ÷
- /
- √

For example:

- 27 ÷ 3
- 27/3
- 3√27

1. **Special Names for Numbers in Division Equation:** Each factor in a division equation is given a distinct name.
   - The dividend payout multiplier and the number of divisions are equal.
   - The quotient is calculated using division.
   - Remainders: Even distribution is not always possible. That might be an extra quantity. That is all that remains (Saleri, 2010).

Let's go through some division characteristics and principles to assist you comprehend this concept better:

- Every positive integer (excluding zero) may be divided by itself to get the number 1 (Skudrzyk, 2012).

As an example: · 7 ÷ 7 = 1; 25 ÷ 25 = 1

- The residual has no set value when divided by zero.

For example, 6 ÷ 0 = undefined 325 ÷ 0 = undefined

- Zero can have just zero divisors.

For example, 0 ÷ 5 = 0; 0 ÷ 100 = 0

- Divide two whole numbers and the result cannot be a whole.

For example, 15 ÷ 2 = 7.5; 20 ÷ 3 = 6.67

In an accurate division (with no remainder), the dividend equals the divisor times the quotient. This criterion can only be satisfied by entire numbers that are not zero.

For example:

If 30 ÷ 5 = 6 then 5 × 6 = 30

If there are three non-zero whole numbers a, b, and c, and b × c = a, then,

a ÷ b = c and a ÷ c = b

For example:

5 × 10 = 50, then,

· 50 ÷ 5 = 10

· 50÷ 10 = 5

2.2.5. Exponentiation (^)

2^3 = 2 × 2 × 2 = 8 is an example. Exponentiation is done in mathematics by using a base and a power. When writing an exponent, the characters b^n, denote the base and power. Say "b (raised) to the (power of) n" to pronounce it. When n is a positive integer, exponentiation is equivalent to multiplying the base by itself repeatedly: The exponent is frequently represented with a superscript number to the right of the base. b^n, is referred to in various contexts as "b to the nth power," "b (raised) to the power of n," "the nth power of b," or simply "b to the nth." Any complex or real exponent is referred to as exponentiation. Matrixes are one of several algebraic structures for which an integer exponentiation may be supplied. Exponentiation is used in many areas, including economics, population development, chemical reaction kinetics, wave dynamics, and public-key cryptography in computer science (Skudrzyk, 2012).

2.2.6. Order of Operations (PEMDAS/BODMAS)

When combining several operations in an expression, use the following order:

- Parentheses/brackets;
- Exponents/orders;
- Multiplication and division (from left to right).

Subtraction (from the left) and addition (from the right) Multiplication comes first in this case:

$5 + 3 \times 2 = 5 + 6 = 11$

The standard mathematical order is BODMAS. To thoroughly address a problem, you must first determine which component to focus on first. Otherwise, you will receive the incorrect result. BODMAS stands for (Saleri, 2010)

- Brackets (any part contained in brackets comes first);
- Order (operations containing powers or square roots);
- Division;
- Multiplication;
- Addition;
- Subtraction.

The following acts are in accordance with BODMAS, PEMDAS, PIDMAS, BIDMAS, and BEMDAS:

Insert the Brackets First: The top rank is shown in brackets. The math should always take precedence. Are there a lot of brackets? Move away from the center and outward. Each pair of brackets should be removed after the computation has yielded a single integer (Skudrzyk, 2012).

After that, Organize/Index: The usage of indices and powers is required for "Order and Index" operations such as cubing and squaring. The math is now on its way.

Multiply and Divide: In a third-order series of operations, the weights of division and multiplication are identical. It is critical to carry out tasks in the same left-to-right order.

A sum of $18 + 6 + 4 + 8$

It Functions from Left to Right: To get adding and subtracting, divide 18 by 8. Again, the same heaviness. As a result, the fourth and final level of operations is addition and subtraction. As the third and fourth stages, divide and multiply are executed concurrently, from left to right, and with equal weight.

2.2.7. Absolute Value (|x|)

For example, $| -5 |$ and $| 5 |$ are both absolute numbers. The absolute value of n expresses how distant it is from zero. An integer's absolute value, or how distant it is from zero, is always a positive number (or zero in the case of zero). In contrast to relative values, which question "in which direction?" rather than "how far?" absolute values always ask, "how far?" We get | 3 | = 3 and | 3 | = 3 since 3 is three places to the right and three positions to the left of zero. Parentheses, brackets, and curly braces are replaced by bars in absolute-value notation. It is critical to use appropriate notation to avoid misunderstandings. Absolute value bars serve a distinct purpose from parentheses. The following example shows that (3)'s absolute value is not equal to +3. The third should be streamlined.

To deal with the absolute value provided by | 3 |, I must first place the absolute value bars in parentheses. To achieve this, I take the positive of the absolute value's innermost portion (also known as the "argument of"):

$-|-3| = -(+3)$

With parentheses, let me now refute:

$-|-3| = -(+3) = -3$

If the integer contained within it ("argument of") was positive, the sign was not changed while calculating the absolute value. When a negative explanation was provided, the "understood" bonus was decreased to "minus." As a result, the problem is unlikely to appear in any of the assignments but will appear in the tests, which is a small nitpick (Skudrzyk, 2012). Variables are used to conceal the sign and value of a number. You can't tell if the "2" or "4" is stored in the x variable, for example. Could you kindly tell me what the absolute value of x is? Because it is unclear from the letter whether the variable has a positive or negative value, both outcomes must be considered (Saleri, 2010).

If x $\geq$ 0, the absolute value remains constant. Assume that x = 2 and that | x | = 2 equals x. The sign is unaffected by whether x is positive or negative:

For $x \geq 0$, $|x| = x$

The absolute value will have a different sign if x is negative. x equals | 4 | = + 4 = (4) = x when x equals 4. Any negative x requires a sign change, thus x 0 becomes x.

For $x < 0$, $|x| = -x$

The "minus" sign before a variable show that the variable's sign has changed from its previous value rather than denoting "a number to the left of zero. "" means that the sign of the original value has been flipped, not that the number is negative.

2.2.8. Negative Numbers

It is possible for a number to be negative. When negative is multiplied by positive, the result is also negative. There is only one example required: 2 (–3) = –6. Because negative numbers lack a zero, it appears before these digits. On a number line, negative numbers are to the left of zero.

1. **How are Negative Numbers Useful in Daily Life?**

Each number has a distinct significance. Include zero and negative numbers as well. Negative numbers are frequently used. Negative numbers are critical in this circumstance.

- Negative values imply higher temperatures. Temperatures as low as – 15°C (59°F) are common in ice areas.
- Credit is represented as a negative number in banking. Negative balances on bank statements indicate that accounts are overdrawn.
- Sea level" refers to the average elevation of the ocean's surface. As a result, results below sea level are perceived adversely.
- Tests and games receive low ratings. Incorrect responses may have a negative impact.

2. **Negative Integers:** All integers are negative numbers. There are no fractions or decimals allowed. Negative numbers include –7 and –10 (Saleri, 2010).
3. **Rules for Negative Numbers:** Negative integers can be employed in operations such as addition, subtraction, multiplication, and division (Skudrzyk, 2012).

- Negative numbers can also be added. –5 + (–1) = –6
- A positive and a negative number are added to calculate the difference. In general, the highest absolute value is displayed first. –9 + 3 = –6
- Adding two positive numbers of results in a negative one. –9 × 2 = –18
- A positive number is obtained by multiplying a negative integer by itself. –6 × –3 =18.
- To generate positive results, divide negative integers by those of the same sign. –56 ÷ –7 = 8
- When negative integers of various signs are split, negative outcomes are obtained. Figures that are negative: –32 ÷ 4 = –8

4. **Adding and Subtracting Negative Numbers:**

Addition of Negative Numbers: The minus sign is used to add two negative numbers together. –7 + (–4) = –11. As a result, the sum of two negative numbers is also negative. A number line is an example of this. The main rule of the number line is, "When adding a negative number, we shift to the left." Because the rule is observed, the solution is –7 + (–4). By subtracting them and applying the sign of the greater absolute value, a positive integer and a negative integer are added. –9 + (5) = –4. The result is –4 because we use the sign of the larger absolute integer. The use of number lines is advantageous in this case. According to the number line's law, every time we add a positive number, we must shift to the right. Apply the formula to the graph of numbers below. We must flip five digits to go from –9 to –4.

Subtraction of Negative Numbers: A negative integer can be subtracted in the same way that a positive integer can be added. According to the Law of Subtraction, the second integer must be added, and the sign must be adjusted. In the first case, the subtraction rule is used to rule out positive values. 5 + (–6) = 5 – 6 = –1. When adding a negative number, such as 5 + (–6), go to the left side of the number line. Moving six digits to the left, beginning with 5, yields –1. Negative multiplication and division

5. **Multiplication and Division of Negative Numbers:** First rule: When the signs of two numbers are opposite each other, the total is negative. (–) × (+) = (–). As a result, adding a positive number to a negative number still yields a negative result. When multiplied by itself, –3 times 6 is –18.

Second Rule: If the signs of the numbers add up, the outcome is favorable. (–) × (–) = (+); (+) × (+) = (+). As a result, multiplying two numbers, whether positive or negative, yields a positive result. –3 × –6 = 18.

Divide Any Positive or Negative Integer in Half (Saleri, 2010): According to the first rule, a negative integer is always divisible by a positive number. (–) ÷ (+) = (–). (–36) ÷ (4) = –9

The second rule argues that by dividing a negative integer by itself, any positive number may be obtained., (–24) ÷ (–4) = 6

6. **Integers with Negative Exponents:** Keep the following two guidelines in mind while working with negative exponents in integers: When a negative integer is multiplied by an even exponent, the result is always positive. For example, 46 may be written as, -4^6 = –4 × –4 × –4 × –4 × –4 × –4 = 4096.

A negative number is always the product of a negative integer multiplied by an odd exponent. $-9^3 = -9 \times -9 \times -9 = -729$

2.2.9. Rational Numbers

Integer fractions with non-zero denominators are referred to as rational numbers. It is rational because every integer is a fraction with the denominator 1. The numbers 1/2, 5/3, –3/4, and 0 are rational. Addition, subtraction, multiplication, and division are operations on rational numbers. A rational number is defined mathematically as p/q, where q is not zero. Non-zero denominators are present in rational fractions. Rational numbers include 1/2, 1/5, 3/4, etc. The number "0" is rational when expressed as 0/1, 0/2, 0/3, etc. But infinite values render fractions like 1/0, 2/0, 3/0, etc., illogical. Here, compare numbers that are reasonable and irrational (Skudrzyk, 2012).

To determine whether a number is realistic, check these conditions.

- p/q, where $q \neq 0$;
- p/q is decimalizable.

Positive, negative, and zero numbers are all acceptable rational numbers. Rational fractions have integer denominators and non-zero numerators. All whole numbers are rational since they can all be expressed as fractions, making them all real numbers.

1. **Rational Standard Numbers:** The rational number is in standard form if the dividend and divisor do not share any factors and the divisor is positive.

12/36 is logical. Due to the divisor and dividend only sharing one element, it may be decreased to 1/3. The usual number is rational number 1/3 (Saleri, 2010).

2. **Positive and Negative Rational Numbers:** Integers p and q are rational numbers. Q must not be zero. Positive and negative numbers are rational. Positive integers for p and q exist for positive rational numbers. If –(p/q), then p or q are negative numbers. $-(p/q) = (-p)/q = p/(-q)$.
3. **Arithmetic Operations on Rational Numbers:** Mathematical operations on integers are essential. Similar procedures may be performed on rational numbers like p/q and s/t.

Addition: The same denominator is required to add p/q and s/t. Therefore, (pt+qs)/qt.

Example: 1/2 + 3/4 = (2+3)/4 = 5/4

To subtract p/q and s/t, first change the denominator to the same value, then do the operation.

1/2 – 3/4 = (2–3)/4 = –1/4

Numerators and denominators of two rational numbers are multiplied when they are multiplied. P/Q multiplied by s/t results in (ps)/(qt).

1/2 x 3/4 = (1×3)/(2×4) = 3/8

- **Division:** pt/qs = (p/q)(s/t).

1/2 = (1×4)/(2×3) = 4/6 = 2/3

4. **Multiplicative Inverse of Rational Numbers:** The reciprocal of a rational integer's fractional representation, p/q, is the multiplicative inverse of that number. For instance, since the multiplicative inverse of 4/7 is 7/4, 4/7 times 7/4 equals 1.
5. **Rational Numbers Properties:** Rational numbers are a subset of real numbers and adhere to all properties of the real number system. Important characteristics of rational numbers include:

- A rational number is produced by multiplying, combining, or subtracting rational numbers.
- A rational number is maintained when the numerator and denominator are divided or multiplied by the same factor.
- A rational number can be added to zero (Skudrzyk, 2012).

2.2.10. Irrational Numbers

Real numbers that are irrational are those that cannot be divided by two integers. Decimal expansions that don't recur or come to an end. 2, 3, and the golden ratio (5 + 1)/2 are examples of irrational numbers. It is possible to estimate but not to compute irrational numbers. Fractionalizing irrational numbers is impossible. It is impossible to define p/q, where p and q are integers, as an irrational number. It is illogical. RQ is used to represent irrational numbers, where RQ stands for "set minus" and the reverse slash sign. The distinction between real and rational numbers is known as R-Q (Saleri, 2010).

1. **Properties of Irrational Numbers:** Irrational numbers adhere to all characteristics of the real number system as subsets of the real numbers. Irrational figures include:

- The sum of irrational and rational numbers is irrational. For example, if x is irrational, y is logical, and x+y is irrational, then z is irrational.
- Any irrational number will result in irrational when multiplied by any non-zero rational number. If xy=z is true, then x=z/y is true, which goes against the notion that x is irrational.
- There could or might not be an LCM for any two irrational numbers.
- Rational addition or multiplication of two irrational integers, such as √2, is possible. Here, √2 is irrational. A rational number is produced when you multiply it twice 2.
- Irrational numbers cannot be multiplied, as contrast to rational numbers.

2. **List of Irrational Numbers:** Golden ratio, Euler's number, and Pi are well-known irrational numbers. The majority of square roots and cube roots are absurd. Contrary to 3, 4 is fair. Since 4 is a perfect square, it is a rational number. There are infinite irrational numbers for any two real numbers. There are countless irrational integers between 1 and 2. Are numerical anomalies real? In mathematics, real numbers are irrational numbers. Ratios of two integers cannot be used to express irrational numbers. Irrational numbers are always produced by imperfect square roots (Saleri, 2010).

Statement: Rationally or irrationally, irrational numbers can accumulate.

Despite being an irrational number, 2 becomes 2 when multiplied by 2.

(i.e.,) $\sqrt{2} \times \sqrt{2} = 2$

We are aware that produces another irrational number when multiplied by.

(i.e.,) $\pi \times \pi = \pi 2$

A rational or irrational number may result from multiplying two irrational integers.

Irrational Numbers Added Together:

Statement: Rational and irrational numbers can be combined (Skudrzyk, 2012). The sum will either be rational or irrational, similar to the product of two irrational numbers.

A prime example of an irrational number is 3√2+ 4√3,

Another illustration is the rational number (3+4√2) + (–4√2),

We must be careful since adding and multiplying two irrational numbers might result in either an irrational or a rational number.

3. **Irrational Number Theorem and Proof**

- **Theorem:** If a^2 (any positive integer) is divisible by p and p is a prime number, then p divides a.
- **Proof:** The positive integer may be expressed as the product of its primes using the Fundamental Theorem of Arithmetic:

$a = p_1 \times p_2 \times p_3 \ldots\ldots \times p_n$ (1)

Equation (1) is squared to give $a^2 = (p_1 \times p_2 \times p_{3\ldots\ldots\ldots} \times p_{n)}\, (p_1 \times p_2 \times p_3 \ldots\ldots\ldots \times p_n)$

$\Rightarrow a^2 = (p_1)^2 \times (p_2)^2 \times (p_3)^2 \ldots \times (p_n)^2$

The prime factorization of a natural number is said to be unique according to the Fundamental Theorem of Arithmetic, with the exception of the factor order. Only the prime factors, a^2 are $p_1, p_{2,}\, p_{3\ldots,}\, p_n$. One of $p_1, p_{2,}\, p_{3\ldots,}\, p_n$. If p is prime and a factor of a^2. So, p will consider a (Saleri, 2010).

As a result, p divides a and a^2.

4. **How to Find an Irrational Number?** Between 2 and 3, look for irrational numbers. 4 and 9 each have square roots of 2 and 3, respectively. Therefore, between 2 and 3, the irrational numbers are 5, 6, 7, and 8. Any two perfect square numbers can also be obtained between irrational numbers.

Another example Suppose. How can we tell if √2 is an irrational number? √2 is a sensible assumption. Based on rational numbers √2 = p/q .... (1) For p and q, which are co-prime integers, q 0 (Skudrzyk, 2012).

Equation (1) is squared to give $2 = p^2/q^2$

$\Rightarrow p^2 = 2\, q^2$ (2)

2 is a prime factor of p if it is a prime factor of p2, according to the preceding theorem.

Thus, the number p is equal to 2c. $p = 2 \times c$,

Equation (3) is solved for this value of p, and the result is $(2c)^2 = 2\, q^2$

$\Rightarrow q^2 = 2c^2$

A prime factor of q2 is also 2, thus. The theorem states that 2 is a prime factor of q as well.

The conclusion above demonstrates that p and q share the prime number 2, contrary to the initial assumption that they are co-primes. This contradiction results from presuming that 2 is logical.

Rational root number 2. We can also demonstrate that p is irrational if p is prime. Prime number p is irrational in any case.

2.2.11. Real Number Line

The real number line, a visual representation of real numbers, has zero at its center and positive and negative values on either side. There is no clear conclusion to the number line. There is a one-to-one correspondence between the dots on a number line and the numbers in the actual world (Saleri, 2010).

Dots in a line stand for digits. Even though the graphic only shows the integers from –3 to 3, the line encompasses all real numbers since it continues forever in both directions. It's used while instructing students how to add and subtract with negative values. In advanced mathematics, the set R of all real numbers is referred to as "the real line." The Euclidean space, which includes the Euclidean line, is one example of a geometric space, as is the real coordinate space, both of dimension one. It might also be a measure space, topological space, metric space, linear continuum, or vector space (affine space). The real line, like real numbers, is denoted by the symbol R (or a bold "R" on the chalkboard). Since it is the first real or Euclidean space, it is often referred to by those designations. While it may look like a horizontal line, the y-axis of a Cartesian coordinate plane really represents a numerical line. One convention is that positive integers are to the right of zero and negative ones are to the left. The line's infinite length in both the positive and negative directions is shown by arrowheads at both ends. A single up arrow indicates a change in measurement kind. In geometry, a line segment has two terminal points, a ray has just one, and an unending line has none. Every route end in infinity. The first number on the number line is greater than the second number, or vice versa. If you take the first number and subtract it from the second, you get the absolute value of the difference. Get rid of this distinction.Therefore, a line segment between 0 and the later number represents its magnitude. Two numbers can be added by "picking up" the distance between 0 and one of the numbers, and then "setting it down" with the 0 end on top of the other number. Combine two numbers, for instance. Since multiplying by 3 is identical to multiplying by 5 + 5 + 5, the range from 0 to 5 should be placed to the right of 5, and then again to the right of the preceding result. Since 15 is the sum of 5 and 3, the process ends at that point. There are three outcomes of length 5 (Skudrzyk, 2012). It is possible to classify as such: The length from 0 to 2 must be moved to the right by the same amount in order to divide 6 by 2. The endpoint that was originally at 0 must now be at 2. The length should be readjusted to the right of where it was previously. As a result, the right end of length 2 may be found in the interval from zero to six. Since 62 = 3, there were three full lengths of two in a six.

2.2.12. Integers

Integers, a subset of real numbers, include the positive and negative versions of whole numbers as well as zero. It's "Z." Zero, one, two, and three are all integers. Integers have the ability to count using addition, subtraction, and multiplication. "Integer" (meaning "whole" or "intact") in Latin is "intact." Integers include both positive and negative numbers since they have no place for fractions or decimals. Zero, a negative infinity, or a positive infinity. Numbers can be subtracted and added to form integers. Zero results from subtracting four fours. 2 3 1 is a negative whole number. It is possible to derive every integer from the counting numbers, resulting in a set that is closed by subtraction (Saleri, 2010).

1. **Types of Integers:**

   Integers can be either positive or negative, but never zero. The middle ground is zero. When zero is written, no positive or negative signs are included. Positive integers are the natural numbers that are greater than zero. Sometimes they'll have a Z+ status. The numbers 1–7 are the positive integers. All integers start at zero Negative numbers are represented by the letter Z. –1, –2, –3, –4, –5, –6, etc.
2. **Addition of Integers:** When adding two numbers with the same sign, add their absolute values together. Absolute values apply to nonnegative integers. The absolute value of –3 is 3. The last digit shares the initial one's sign.

- 7+3=10

The numbers 3 and 7 are both nice. The answer is either +10 or 10.

- (–7) +(–3) =–10

Seven and three are two absolute numbers that add up to 10. Since both numbers are integers and contain a dash (–), the answer is –10. When adding integers with opposite signs, you subtract the absolute value of the smaller number. The sign of the number should correspond to its largest absolute value. –4+8 =+4. Here, both 4 and 8 are being treated as absolute values. Counting backwards, we get 4. The largest absolute value is eight (+). It's +4. –6 + 2 = –4. Here, 6 and 2 have the same value in absolute terms. Counting backwards, we get 4. A minus sign can only represent absolute values up to a maximum of six. Subtracting integers yields a negative result of –4 (Skudrzyk, 2012).

3. **Subtraction of Integers:** You may add by switching the sign of the second integer in the subtraction. You can solve these using techniques from integer addition.

(–6) – (+7) = (–6) + (–7) =13.

To start, 7 is multiplied by its new sign. To add, you must follow the guidelines. Add up the digits: 6+7 = 13. The answer is –13 since there is a (–) in each.

4. **Multiplication of Integers:** If the signs of the two numbers being multiplied are the same, then the outcome of the multiplication will be a positive (+) sign, otherwise it will be a negative (–) sign (Saleri, 2010).

Division of Integers: The division of two numbers with the same sign will produce a positive (+) or negative (–) result.

5. **Properties of Integers:**

- The closure property states that operations on integers, such as multiplication and addition, always result in integers. For any pair of integers, a and b:
- a + b = Integer
- a – b = Integer
- a × b = Integer

Since integer division can provide non-integer results, the closure feature is useless. 15÷2=7.5

Commutative Property: The commutative property states that while adding or multiplying, changing the integers has no effect. This quality has no effect on dividing or subtracting. –5 + 2 = 2 + (–5) = –3

Associative Property: According to the associative property, the results of operations are unaffected by how the numbers are grouped. As previously, division and subtraction are not valid operations.

Distributive Property: The distributive property simplifies multiplication by reducing it to simple addition and subtraction (Skudrzyk, 2012).

- $5 \times (4 + 3) = (5 \times 4 + 5 \times 3) = 35$
- $-2 \times (6 + 1) = (-\$\$2) \times 6 + (-\$\$2) \times 1 = -14$

Additive Inverse Property: When adding a negative integer value, you always get zero.

- $6 + (-6) = 0$;
- $15 + (-15) = 0$;

Multiplicative Inverse Property: Multiplication of an integer by its reciprocal yields 1.

- $3 \times 13 = 1$
- $-15 \times 1 - 15 = 1$

Integer Identity: The sum of an integer and zero is always zero. If you multiply any number by 1, you get back the same integer.

If you add zero to ten, you get ten.

- $10 + 0 = 0 + 10 = 10$
- $10 \times 1 = 1 \times 10 = 10$

2.2.13. Prime Numbers

Prime numbers are those natural numbers greater than one that can be divided only by themselves and 1 without resulting in a negative number. The values here are between 2 and 13. Numerology and encryption both rely on prime numbers. Prime numbers are defined as natural numbers greater than 1 that cannot be written as the sum of two smaller natural numbers. Composite numbers are those that are not prime. The number 4 is composite because it is the product of two even smaller numbers (2 x 2), while the number 5 is prime since it produces just 1 and 5. Because every natural number greater than one is either a prime or can be factored as a unique product of primes, primes play a significant role in number theory (Skudrzyk, 2012).

Only 1 and the prime number itself make up the set of positive divisors. Only between itself and one other person is it possible to divide it fairly. Only one and five can be used to divide 5. Because it may be evenly divided by 1, 2, 3, and 6, the number 6 is not prime. The digits 2, 3, 5, 7, 11, 13, 17, 19, etc., are all prime numbers. Given their versatility, prime numbers serve as the building blocks from which all other natural numbers are constructed.

This is the "Fundamental Theorem of Arithmetic."

Prime numbers were initially examined by Euclid and Eratosthenes. In his work "Elements," Euclid proved that prime numbers could go on forever. The prime number distribution is a standard tool in mathematics. Prime numbers become scarcer when compounded. Prime numbers are created at random. Mathematicians have spent years studying challenging mathematical ideas like the Riemann Hypothesis, which has yet to be verified. The use of prime numbers extends well beyond mathematics. They are essential for digital communication and cryptography. Encryption keys for sensitive information on the Internet are often large prime integers. The difficulty of factoring two huge prime numbers into their primes is used in cryptography processes (Saleri, 2010).

The RSA algorithm relies on the product of two extremely large prime numbers that are mathematically impossible to factor with present technology. Prime number conversations are safe using RSA encryption. The ability to compute allows for the discovery of primes that are both bigger and more complex. Distributed computing methods find really large prime numbers with the help of people all around the world.

2.2.14. Least Common Multiple (LCM) and Greatest Common Divisor (GCD)

The least common multiple (LCM) of any two or more integers is that number. The greatest common divisor (GCD) of a pair of numbers is one. The GCD of any two or more integers will equally and completely divide the numbers you specify. Using the LCM and GCD can simplify calculations and fractions (Skudrzyk, 2012).

2.2.15. Inequalities

Inequalities compare two real numbers or expressions using symbols like (less than), >, (greater than), and (less than or equal to). Value increases are indicated by $X > 3$, whereas decreases are shown by Y 7 (Saleri, 2010).

CHAPTER 3

Fundamental Mathematical Objects

Contents

A mathematical object is an abstract idea that is formed inside the field of mathematics. A mathematical object is anything that has a formal definition and may be used in the deductive reasoning and mathematical proofs that are connected with mathematics. A mathematical object is often nothing more than a value that may be used in a calculation, however there are certain exceptions to this norm. Numbers, sets, functions, expressions, geometry, transformations, and spaces are common examples of items found in mathematics. In the domain of proof theory, theorems, proofs, and even hypotheses may be considered mathematical objects. Mathematical concepts are often difficult to grasp. Mathematical philosophers have spent a great amount of time and effort investigating and discussing the ontological quality of mathematical objects. Some of the fundamental mathematical objects are discussed as follows (Ong, 2020):

3.1. NUMBER THEORY

The fundamental topic of number theory, sometimes known as conventional mathematics or advanced mathematics, is the study of integers and arithmetic functions. Carl Friedrich Gauss (1777–1855) is quoted as saying, "Mathematics is the queen of the sciences, and number theory is the queen of mathematics." Number theory is the study of prime numbers and the qualities of mathematical objects that are generalizations of integers or are derived from integers (such as rational numbers or algebraic integers). The discipline of Diophantine geometry provides a framework for seeing integers as both equations and solutions. Number-theoretic concerns are often best understood by using analytic objects (such as the Riemann zeta function) that incorporate features of integers, primes, and other number-theoretic objects. This is usually the most basic and easy strategy. The rational number approximation devised by Diophantus allows them to be compared to real numbers. Number theory, a more conventional phrase for mathematical number theory, is often used. At the turn of the 20th century, the field of study known as "number theory" was already well-established. The word "arithmetic" has taken on new implications in the realms of mathematical logic and computer technology, such as floating-point arithmetic, which is equivalent to "elementary calculations" in common parlance. Another case in point is the discipline of mathematical logic, where the word "arithmetic" has acquired new connotations. The word "arithmetic" gained substantial momentum in the field of number theory in the second half of the 20th century, perhaps as a result of the influence of French culture on the science.

Mathematical approaches are not always preferred over those based on number theory (Ong, 2020).

3.1.1. Elementary Number Theorem

In most circumstances, a simple operation does not need a complex analysis. Although sophisticated analysis was used to show the existence of prime numbers in 1896, it wasn't until Erds and Selberg in 1949 that a plain proof was achieved. Even when Fourier analysis is used rather than complex analysis, demonstrations relying on advanced Tauberian theorems (such as Wiener-Ikehara) are often viewed as interesting but not essential. This is true even when Fourier analysis is used. As is frequently the case in other circumstances, the great majority of readers may find even the most basic evidence confusing and difficult to comprehend. In 1985, a number theorist called Paul Erds was 72 years old, and a number theorist named Terence Tao was 10 years old (Nathan, 2012). It is a prevalent misperception that laypeople can understand a major percentage of number theory breakthroughs. The employment of a wide variety of very complex mathematical procedures, on the other hand, suggests that the explanations for these findings are sometimes, at best, hazy.

3.1.2. Analytic Number Theory

The Riemann zeta function for variables having complex values. The color of a point represents the point's argument, with deeper colors representing values closer to zero. The upper half of the plane exhibits modal group dynamics. The grey region indicates the most basic characteristics of common ground. Analytic number theory may be thought of as the study of size and density estimates rather than number theory identities, or as the study of integers using real and complex analytic methods. In any case, think of it as the study of number theory. When it comes to capturing some topics in analytic number theory, such as sieve theory, the second formulation is more appropriate. Despite the fact that certain components of this theory only use analysis on the surface, they are nonetheless called analytic number theory (Ong, 2020).

Questions in analytic number theory include the prime number theorem, the Goldbach conjecture (also known as the twin prime conjecture, the Hardy-Littlewood conjectures, or the Waring issue), the Riemann hypothesis, and the Waring problem. Some of the most fundamental approaches in analytical number theory include the circle technique, the sieve method, and L-functions (or, more precisely, the study of the properties of these functions). The study of modular forms, as well as automorphic forms in general, has lately gained prominence in the field of analytical number theory.

At the confluence of algebraic number theory and analytic number theory, the latter may be utilized to address the former's worries about the former. Prime ideals, which are generalizations of prime numbers that originate in algebraic numbers, may be defined. Furthermore, the number of prime ideals up to a specific size may be queried. The Riemann zeta function is a fundamental analytic object at the heart of the problem, and it may be handled by using Dedekind zeta functions, which are generalizations of the Riemann zeta function. We may learn about the distribution of a sequence (in this example, prime ideals or prime numbers) by assessing the behavior of a well-built complex-valued function (Nathan, 2012).

3.1.3. Algebraic Number Theory

For a number to be termed an algebraic number, it must satisfy the equation f (x) = 0 f(x)=0. Algebraic numbers must have complex numbers with rational coefficients. Solutions to equations like x = 0 may be written as algebraic integers in the following format: x 5 + (11/2) x 3 7 x 2 + 9. The term "number fields" refers to a broad idea rather than to real or algebraic numbers. The emphasis of algebraic number theory study is on number fields that only include algebraic numbers. As a consequence, there is often duplication of effort between analytical and algebraic number theory, the former distinguishable from the latter by the methods it utilizes and the latter by the themes it investigates (Ong, 2020).

Given that Gauss's analysis of quadratic forms in Disquisitiones arithmeticae may be recast in terms of ideals and norms in quadratic fields, it seems conceivable that he previously investigated the simplest feasible kind of number field. Because ideals and norms in quadratic fields may be recast in terms of ideals in quadratic fields, this is the case. (The values for a, b, and d in a quadratic field are all fixed rational integers, while the values for the square roots of a, b, and d are not.) Furthermore, the chakravala approach, which originated in India in the 11th century and is still in use today, might be regarded a modern way for determining the units of a real quadratic number field. However, Gauss and Bhskara were not familiar with the concept of number fields at the time (Nathan, 2012).

The contemporary discipline of mathematics may have its roots in the late 19th century, when three independent but related approaches to the problem of non-unique factorization in algebraic number fields were developed. The

Riemann hypothesis, the factorization triangle, and the factorization matrix are all names for these procedures. The debate included ideal numbers, the theory of ideals, and value theory. (For example, in the field generated by the rationals and 5 sqrt –5, the number 6 6 may be factored as 6 = 2 3 6=2cdot 3, 6 = (1 + 5) (1 5) 6=(1 + sqrt –5)(1 – sqrt –5), and so on, for a grand total of 2, 3, 1, 5, and 1 + sqrt –5. Kummer's interest in ideal numbers may have been sparked by generalizations of quadratic reciprocity or other laws of higher reciprocity (Nathan, 2012).

Extensions of fields with a lower number are often studied; if field L encompasses field K, then field L is an extension of field K. The complex number C may be extended in a variety of ways, most notably via the reals R and rationals Q. How to identify hypothetical expansions of a certain number field is a difficult issue, but one with plenty of potential for investigation. The term "L of K extension" is often used to describe an abelian extension. This is due to the fact that the Galois group Gal(L/K) of L over K is abelian. Class field theory was an effort to categorize these occurrences that began in the late 1800s (in part by Kronecker and Eisenstein) and was mostly completed between 1900 and 1950. One of the subjects treated in modern algebraic number theory is Iwasawa theory. A typical definition of the Langlands project is the research of non-abelian extensions of number fields using class field theory. This is one of the most significant mathematical research projects presently underway (Ong, 2020).

3.1.4. Diophantine Geometry

The fundamental topic of Diophantine geometry is whether or not a given Diophantine equation has solutions, and if so, how many solutions that equation has. Consider equation solutions to be distinct shapes that may be formed using geometry. For example, if you wish to construct a planar curve, you may use a two-variable equation. A curve, surface, or other equivalent object in n-dimensional space may be defined by an equation or sequence of equations containing two or more variables. One of the most fundamental concerns in Diophantine geometry is whether or not a curve or surface contains rational points, which are points with all rational coordinates. If such regions exist, the next step is to determine how many people dwell there and how they are spread over the land. One of the most difficult components of this approach is determining if a curve (or surface) has a limited or infinite number of rational points (Nathan, 2012). There is a lot of interest in finding solutions to the Pythagorean equation (x 2 + y 2

= 1) that explain both x and y. Because each answer to the second equation yields a solution to the first equation, x = a/c and y = b/c, requesting all integer answers to the second equation is analogous to requesting all integer replies to the first equation. It is equivalent to requesting rational coordinates for all points on the curve x 2 + y 2 = 1, which is the same thing. This curve generates a circle with a radius of one and a center point at the origin (Ong, 2020).

The diagram below depicts two separate genus-one elliptic curves, each with at least one rational point. (One way to think of each graph is as a four-dimensional part of a torus.)

It has come to our notice that plotting points on curves as a method of solving equation problems might be useful. A curve's genus is the number of rational or integer points on the curve that are solutions to the equation f (x, y) = 0 f(x, y)=0, where f f is a two-variable polynomial. The genus of a curve determines the number of rational or integer points on it. The term "genus" will be defined later in this chapter. The equation f (x, y) = 0 may be used to represent a two-dimensional surface in a (projective) four-dimensional space since two complex variables can be decomposed into four real variables. The genus of the equation f (x, y) = 0, which equals zero, is the number of holes on the surface that are formed like doughnuts. Other geometrical notions, it turns out, are just as significant.

Another area of study related to this issue is diophantine approximations. These computations show how near rationals may go to a given number (x x). (We need exact estimates for the amount of space needed to record the rational: If the ratio | x a/q | 1 q c |x-a/q|frac 1qc is big, then the approximation a/q a/q of x x (with gcd (a,q) = 1 gcd(a,q)=1) is correct. This is the only prerequisite for it to be legitimate. If x x is a number that can be expressed algebraically, the subject takes on a whole new level of complexity. Many equations cannot have integer or rational solutions if the values of x and x are not valid approximations. Furthermore, some ideas, the most essential of which is height, are necessary to understand Diophantine geometry and Diophantine approximations. This is especially important in the study of transcendental numbers since a transcendental number may be approximated with more accuracy than any algebraic number. This line of reasoning was used in order to highlight and e's metaphysical stance (Nathan, 2012).

Number geometry, on the other hand, is a set of graphical tools that aids in the solution of algebraic number theory issues. Number geometry is not the same as diophantine geometry. The word "arithmetic geometry" relates to diophantine geometry, an ancient kind of geometry that has been around for a long time. Unlike the procedures employed in Diophantine approximations, the word "arithmetic geometry" is more likely to be used when referring to modern algebraic geometry (as in Faltings' theorem).

The following issues are based on verifiable material, although none of them are older than the mid-20th century. Researchers did not begin investigating the notion of computability until the 1930s and 1940s, and the concept of computational complexity was not thought of until the 1970s. For example, the problem of algorithms in number theory dates back far before the idea of proof was even thought of (Ong, 2020).

3.1.5. Probabilistic Number Theory

Probabilistic number theory may be classified into an important area that focuses on the analysis of variables that are significantly related but not totally independent of one another. The fact that an integer between 1 and 1 million is divisible by two has almost little bearing on whether or not it is also divisible by three.

The assumption that occurrences with probability larger than zero must occur on occasion is a cornerstone of probabilistic combinatorics, and it is used in many applications of probabilistic number theory as well. This is a widely believed concept that is backed by a substantial amount of evidence. This is an example of a very particular non-probabilistic argument originating from a probabilistic one; if certain algebraic objects can be demonstrated to be at the tail of certain rationally specified distributions, then these things must be rare. These items contain rational or integer answers to certain equations. Cramer's hypothesis is only one example of how a probabilistic technique with erroneous parameters may generate more problems than it solves (Ong, 2020).

3.1.6. Arithmetic Combinatorics

The Fundamental Ingredients Both combinatorics and number theory have a strong mathematical focus. Would the growth of a high number of the arithmetic progression's components arise from starting with a particularly "thick" infinite set A A? Among them are the letters "a," "a+b," "a+2," "a+3," and "a+10," to name a few. When A and A are joined, it is expected that they would be able to represent numbers of the highest order. Issues like this often arise as difficulties in the subject of arithmetic combinatorics.

This field includes additive number theory (which deals with very particular sets A A of mathematical importance, such as the primes or squares), maybe some number geometry, and rapidly expanding research. Additive number theory is concerned with highly specific sets A A of mathematical relevance, such as primes and squares. Its growing connections to other areas, such as ergodic theory, finite group theory, model theory, and others, may be explained in part by its emphasis on growth and distribution issues. The sets A A under consideration, on the other hand, may not have to be integers. They may also be subsets of non-commutative groups, in which case the sign for multiplication is used instead of the symbol for addition, or rings, in which case the expansions A + A A+A and A AA A A can be compared (Nathan, 2012).

3.2. COMBINATORICS

Mathematical combinatorics use the characteristics of finite structures as well as counting to arrive at its conclusions. It has practical applications in subjects as diverse as computer science, statistical physics, logic, and evolutionary biology, and it is connected to a broad range of mathematical subfields. Combinatorics, as a branch of study, is generally recognized as holding the key to addressing a broad range of problems. Combinatorial problems may be encountered in a broad range of mathematical subfields, including algebra, probability theory, topology, and geometry, to name a few. Many combinatorial problems have been addressed in the past by employing a short-term solution to a problem that was initially raised in a specific discipline of mathematics. By the end of the 20th century, however, strong and comprehensive theoretical procedures had been created, and combinatorics had evolved into a distinct mathematical subject capable of standing on its own. Graph theory is an important topic of study in the subject of combinatorics, with various evident linkages to other academic subfields. Combinatorics' equations and estimations are frequently employed in the process of algorithm analysis (Ong, 2020).

Combinatorialists are a subgroup of mathematicians who specialize in the study of combinatorics, a subject of mathematics.

Combinatorics is sometimes undervalued due to its broad variety of applications. According to H. J. Ryser, the issue cannot be categorized since it combines aspects from a broad range of mathematical subfields. The following is a list of several types of issues that combinatorics aims to solve. Structure enumeration (finding the "best" structure or solution among multiple possibilities, whether "largest," "smallest," or satisfying some other optimality criterion), structure construction (possibly in multiple ways), structure optimization (finding the "best" structure or solution among multiple possibilities, whether "largest," "smallest," or satisfying some other optimality criterion) (Nathan, 2012).

According to Leon Mirsky, combinatorics is a group of domains that explore the same thing but have wildly disparate aims, approaches, and degrees of coherence. Combinatorics is the umbrella term for several disciplines. Combinatorics may be characterized in a variety of ways, one of which is by characterizing its subfields and the challenges and opportunities they present. The process is described in detail here. Looking back in time, the inclusion or removal of some disciplines under combinatorics may be justified. Despite the fact that the great majority of combinatorial challenges and methodologies are focused on finite systems, they may be applied to an infinite (or, more precisely, countable) setting that is still discrete (Ong, 2020).

Both enumerative and combinatorial mathematics are said to have originated in the ancient world. According to the sixth-century BCE Indian physician Sushruta, who published the book Sushruta Samhita, it is feasible to calculate all 26 tastes by mixing six unique flavors in any of 63 different ways (one at a time, two at a time, etc.). This is mentioned in the book. According to the Greek historian Plutarch, Hipparchus, a mathematician, and Chrysippus, a philosopher, fought about an enumerative issue known as the Schroder-Hipparus numbers around the second century BCE. In the third century BCE, Archimedes may have examined the number of different configurations for a tiling assignment, while Apollonius' lost writings may have addressed combinatorial issues.

During the Middle Ages, academics from civilizations other than Europe were in charge of studying combinatorics. Although Indian mathematicians may have recognized the equations for the number of permutations and combinations as early as the sixth century C.E., these formulas were not codified until the time of Mahvra (about 850). Rabbi Abraham ibn Ezra, an astronomer and philosopher, was the first to show the symmetry of binomial coefficients in 1140. Levi ben Gerson (also known as Gersonides) developed a closed formula based on the symmetry of binomial coefficients in the year 1321. As a pictorial representation of the relationships between binomial coefficients, mathematicians in the 10th century created the arithmetical triangle, often known as Pascal's triangle. Later, in the Middle

Ages, campanology offered Cayley graphs on permutations, allowing the discovery of several instances of what are now known as Hamiltonian cycles (Ong, 2020).

During the Renaissance, the discipline of combinatorics, like many other subfields of mathematics and scientific investigation, saw a renaissance. Pascal, Newton, Jacob Bernoulli, and Euler all made substantial contributions to the extension and development of science in their early works. In the early 20th and late 19th centuries, Percy MacMahon and J. J. Sylvester set the basis for what is now known as contemporary enumerative and algebraic combinatorics. During this time, the four-color issue revived interest in graph theory. Combinatorics grew in popularity rapidly in the second half of the 20th century, resulting in the formation of new publications and conferences in the area. The discovery of new links and applications to surrounding subjects like as algebra, probability, functional analysis, number theory, and so on contributed significantly to the field's growth. These links have helped to blur the boundaries between the different subfields of mathematics and theoretical computer science, but they have also contributed to the field's rising fragmentation (Nathan, 2012).

3.3. SET THEORY

Set theory is a branch of mathematical logic that studies groups or collections of various things. Despite the fact that everything may be put into a set, the major emphasis of the mathematical field known as set theory is on mathematical concerns. In the 1870s, two German mathematicians, Richard Dedekind and Georg Cantor, lay the basis for what is now known as contemporary set theory. It is widely acknowledged that Georg Cantor was the first to create set theory. A "naive set theory" is a preliminary study of non-formalized systems. At the start of the 20th century, problems in naive set theory led the creation of a number of distinct axiomatic systems. Russell's dilemma, Cantor's paradox, and the Burali-Forti conundrum are examples of axiomatic systems.

By far the most well-known and intensively explored set theory is the Zermelo-Fraenkel set theory (ZFC; with or without the axiom of choice). Because this has been argued on occasion, it is feasible that all of mathematics may be traced back to the ZFC and the axiom of choice. Set theory has a wide range of applications, including formal semantics, computer science (particularly relational algebra theory), philosophy, and other disciplines of study. Because of its intuitive appeal, paradoxes, and implications for the idea of infinite, set theory is a subject of great interest to logicians and philosophers of mathematics. It also has a wide variety of uses. One of the subjects being researched by current set theorists is the stability of very large cardinal numbers. Other issues being studied include the geometry of the real number line (Ong, 2020).

3.3.1. Ontology

A set is said to be "pure" if each and every component that makes up the set is itself a set. A nonempty pure set is represented as a set with just itself as its members. Many axiomatic set theory systems are built to axiomatize only pure sets, and modern set theory often concentrates on the von Neumann world of pure sets. Because practically every mathematical notion can be described in terms of a pure set, this constraint results in huge practical improvements at the cost of merely a smidgeon of generality. In the von Neumann world, sets are structured into a cumulative hierarchy, with the depth of nesting of their members, members of members, and so on being the key deciding factor. Each set in this hierarchy is assigned the ordinal number alpha, also known as the rank, using transfinite recursion as the mechanism of assignment. A pure set's rank, given by X X, is the ordinal that is the least stringently larger than any of the set's components. An empty set has a rank of zero, but a set that includes only empty sets has a rank of one. Set V V_alpha is a collection of all pure sets with a rank less than the ordinal alpha. This definition is true for any ordinal alpha. Von Neumann's whole life may be summed up in two letters: V V.

3.3.2. Axiomatic Standpoint

The ideas of set theory, which may be taught intuitively and informally, can be conveyed to primary school students by their teachers using Venn diagrams. The intuitive approach assumes, rather than declares explicitly, that it is possible to construct a set from the group of all items that fulfil a certain condition. This assumption generates a number of paradoxes, the most basic and well-known of which are Russell's dilemma and the Burali-Forti conundrum. As a consequence of this assumption, more paradoxes emerge. Axiomatic set theory was initially presented as a solution to these difficulties with set theory. Each set is seen as a component in a hierarchical structure that may be characterized as cumulative by the most well-known axiomatic set theory methods. Depending on the ontologies they employ, these systems may be classified as one or both of the following:

Different preparations were undertaken. The most well-known example of an axiomatic set theory is the ZFC, which integrates the axiom of choice. The Zermelo set theory, the General set theory, the Kripke-Platek set theory (which removes the infinity, powerset, and choice axioms and weakens the separation and replacement axiom schemata), and the Peano axioms and finite sets are all examples of ZFC fragments. There are several types of meetings and assemblages that are legal. When it comes to the strength of theorems regarding sets on their own, ZFC performs worse than Morse-Kelley set theory, Tarski-Grothendieck set theory, and Von Neumann-Bernays-Godel set theory. Such systems may use urelements, which are defined as things that are neither empty nor part of a set (Ong, 2020).

Van Orman, Willard Quine's New Foundations systems, NFU (which allows urelements) and NF (which does not), do not have a cumulative hierarchy. The NFU system permits urelements, but the NF system does not. In both NF and NFU, the "set of everything" comprises all of the missing complements for the other sets. In these cases, urelements play an important role since the choice axiom does not hold true for sets generated by NF but does hold true for sets made by NFU. Despite the fact that it does not explain the usual cumulative hierarchy and fails the well-foundedness requirement, Thomas Forster claims that the NF ontology is illustrative of an iterative understanding of set (Nathan, 2012).

In contrast to classical logic, constructive set theory frameworks such as CST, CZF, and IZF establish set axioms in terms of intuitive understandings of the concepts involved. On the other hand, there are systems that employ conventional logic but have an unusual membership link. According to these ideas, the value of an atomic formula encoding a membership relation may be anything other than True or False. Fuzzy set theory and rough set theory are two examples of these theories. ZFC models with Boolean values are linked together. Edward Nelson presented internal set theory in 1977 as an extension of ZFC.

Set theory notions may provide a detailed account of many other fields of mathematics on their own. Axiomatically rich sets may be used to build mathematical structures such as graphs, manifolds, rings, vector spaces, and relational algebras. Vector spaces may also be used to create these structures. Set theory may be used to construct mathematical concepts that incorporate connections, such as equivalence and order relations. Set theory is another exciting discovery in the world of mathematics. Since the first book of Principia Mathematica was released, it has been said that each mathematical theorem may be deduced from a correctly formulated set of axioms from set theory, as well as numerous definitions using first- or second-order logic. This has been the case since the publication of the first book of Principia Mathematica. Set theory may be used to discover the characteristics of natural and real numbers, for example, since every number system has an associated set of equivalence classes under a suitable equivalence relation with an infinite field. This is conceivable because the field of a set can never be exhausted (Nathan, 2012).

It is commonly believed, at least in principle, that theorems in mathematical analysis, topology, abstract algebra, and discrete mathematics may be deduced from set theory axioms and definitions. Set theory serves as the basis for a broad range of mathematical subdisciplines. However, only a few key set theory theorems have been proven in a formal context. This is because the derivations necessary to establish these theorems are sometimes significantly more time-consuming than the plain language proofs generally provided by mathematicians. The ZFC set theory, first-order logic, and propositional logic served as the foundation for the study, which culminated in the computer proving over 12,000 theorems in the Meta-math language (Ong, 2020).

3.4. GEOMETRY

Geometry, a discipline of mathematics, deals with themes such as distance, shape, size, and the relative arrangement of distinct things. It is derived from the Ancient Greek phrases "gemetra" (meaning "land measurement"), "gê" (meaning "earth, land," and "v" (metron "a measure"). Geometry, like arithmetic, is a cornerstone of the mathematical canon. A geographer is a kind of mathematician who is a specialist in geometry. Point, line, plane, distance, angle, surface, and curve are all key concepts in Euclidean geometry, which before the 19th century. While the initial goal of geometry was to depict the physical universe, it is currently employed not just in the sciences but also in the arts, architecture, and other related professions. Geometry has the potential to assist many seemingly unconnected areas of mathematics. Wiles extensively use algebraic geometry in his presentation of Fermat's Last Theorem, which was first given in terms of basic arithmetic but remained unsolvable for decades.

Numerous discoveries made in the 19th century considerably expanded the scope of geometry. Carl Friedrich Gauss made one of the earliest such discoveries with his Theorema Egregium ("remarkable theorem"), which argues that the Gaussian curvature of a surface is independent of any particular embedding in a Euclidean space. This means that surfaces may be studied intrinsically or as independent spaces, and the concept has been extended to manifold theory and Riemannian geometry. It became obvious in the late 19th century that non-Euclidean geometries might be constructed without the parallel postulate. Because general relativity is based on non-Euclidean geometry, it is one of the most well-known applications of this kind of geometry (Ong, 2020).

Differential geometry, algebraic geometry, computational geometry, algebraic topology, discrete geometry, often known as combinatorial geometry, and other fields of geometry have emerged since the late 1800s. Another example is affine geometry, which solely concerns point alignment while ignoring distance and parallax. The growth of geometry has resulted in a change in the meaning of the word "space," which formerly referred primarily to the three-dimensional space of the physical world and the Euclidean geometry paradigm. As it stands, "geometric space" or "space" in general refers to a mathematical framework that determines a certain geometry (Nathan, 2012).

3.4.1. Axioms

One of the most important texts ever published, Euclid's Elements, presented geometry in an abstract form. Euclid's axioms and postulates represent the inherent or self-evident qualities of points, lines, and planes. Then he used math to accurately deduce other traits. Because of this distinguishing feature, Euclid's approach of learning geometry has been referred to as axiomatic geometry or synthetic geometry. Non-Euclidean geometries were found in the early 19th century by Nikolai Ivanovich Lobachevsky (1792–1856), János Bolyai (1802–1860), and Carl Friedrich Gauss (1777–1855), prompting fresh interest in the subject. In the 20th century, David Hilbert (1862–1943) used axiomatic reasoning to build a new foundation for geometry (Ong, 2020).

3.4.2. Points

Points are widely considered as geometry's most fundamental unit. They may be defined using synthetic geometry or required features, similar to how Euclid defined "that which has no part." They are often viewed as constituents of the axiomatically defined set known as space in modern mathematics. A line, rather than the collection of points it links, is a fundamental element in synthetic geometry. In current formulations, any geometric shape is reduced to a collection of points. Points are not the essential building elements of contemporary geometry. Point-free geometry, invented by Alfred North Whitehead in 1919–1920, was one of the first of these geometries (Nathan, 2012).

3.4.3. Lines

According to Euclid, lines "lie equally with respect to the points on themselves" and have "no breadth." Given the numerous distinct geometries that contemporary mathematics may depict, the concept of a line is inherent in the representation itself.

In the discipline of analytic geometry, a line in the plane is often defined as the set of all points whose coordinates satisfy a certain linear equation. In a broader context, such as incidence geometry, a line may exist irrespective of the collection of points it links. A geodesic is a differential geometry concept that extends the concept of a line to curves.

3.4.4. Planes

In Euclidean geometry, a plane is a flat, two-dimensional surface having an unlimited extent; other forms of geometry are based on generalizations of this concept.

Planes have various uses in geometry. Planes, for example, may be treated as topological surfaces that are not constrained by lengths or angles. It may be analyzed as a complex plane using complex analytic techniques, or as an affine space, in which collinearity and ratios can be evaluated but not distances (Nathan, 2012).

3.4.5. Angles

Euclid defined plane angles as the intersection of two nonperpendicular lines in a plane. An angle is now defined as the geometric shape formed by two beams meeting at a common terminal. Acute (a), obtuse (b), and right (c) angles are the three types of angles. Oblique angles are made up of both acute and obtuse angles. Angles are studied in Euclidean geometry both independently and in relation to polygons and triangles. Trigonometry is founded on the study of angles in triangles and unit circles. The derivative is used in differential geometry and calculus to calculate the angle between a plane curve, a space curve, and a surface.

3.4.6. Curves

A curve, like a line, is a one-dimensional element that may be straight or curved. Curves have three dimensions in space, but only two on a flat surface. A curve in topology is a function that maps one space to another. Differential geometry employs the same approach, but the defining function must be differentiable. The emphasis of algebraic geometry is algebraic curves, which are one-dimensional representations of algebra. In two dimensions, both spherical and parabola-shaped surfaces may be represented. Surfaces are described in two dimensions via diffeomorphisms or homeomorphisms in differential geometry or topology as "patches" (or neighborhoods). In algebraic geometry, polynomial equations are used to characterize surfaces (Ong, 2020).

When you combine the concepts of curve and surface, you obtain manifolds. A manifold is a topological space in which every point is surrounded by other points in Euclidean space that are similarly homomorphic to each other. A differentiable manifold is a space in differential geometry in which every neighborhood has a differentiable connection to the underlying metric space, Euclidean space. In physics, manifolds have several applications, notably in string theory and general relativity (Nathan, 2012).

3.4.7. Length, Breadth, and Height Measurements

The length, area, and volume of an item, in that order, determine its one-dimensional, two-dimensional, and three-dimensional dimensions. In both analytic and Euclidean geometry, the Pythagorean theorem may be used to calculate the length of a line segment. Area and volume may be defined, described, and computed in terms of lengths in a two- or three-dimensional space, but they are essentially distinct notions. Mathematicians have developed a broad range of exact equations for calculating the area and volume of various geometric objects. Integrals used in calculus to calculate areas and volumes include the Riemann integral and the Lebesgue integral.

3.4.8. Dimensions

The metric system abstracted the concept of length or distance. For example, the Euclidean metric is used to estimate the distance between Euclidean plane points, while the hyperbolic metric is used to find the distance between hyperbolic plane points. The Lorentz metric of special relativity and the semi-Riemannian metric of general relativity are two more prominent metrics. Measure theory cleverly broadens the meaning of length, area, and volume by investigating various methods of providing a numerical value to each of these concepts.

3.4.9. Similarity and Consistency

When two shapes have the same dimensions and angle ratios, we say they are congruent or similar. Two shapes are considered to be congruent in Euclidean geometry if they are the same size and shape. In his attempts to offer a more formal foundation for geometry, Hilbert considered congruence as an indeterminate word whose features are specified by axioms. Congruence and resemblance are wide concepts in transformation geometry, which is the study of the properties of geometric objects preserved by various forms of transformation. The construction of geometric objects was given a lot of study in classical geometry. Traditionally, most geometric drawings were created with just a compass and a straightedge. Furthermore, each construction project must go through a certain number of phases. While these tactics were beneficial in many cases, others necessitated more imaginative approaches, such as the employment of mechanical devices and curves such as Nuesis and parabolas (Ong, 2020).

In the past couple 100 years or more, mathematicians and physicists have increasingly turned to higher dimensions in lieu of the three dimensions utilized in classical geometry—a line, a plane, and our whole cosmos. Consider the configuration space of a physical system, which has the same dimension as the system's degrees of freedom, as an example of a practical application of higher-dimensional mathematics. Five coordinates, for example, would be sufficient to explain the fastener's design. Dimension in general topology now includes both infinite and positive real numbers (similar to fractal geometry and Hilbert spaces). In algebraic geometry, there are several definitions of an algebraic variety's dimension, and they are all equal in most circumstances (Nathan, 2012).

3.4.10. Symmetry

The concept of symmetry dates back virtually to the beginning of formal geometry study. The circle, regular polygons, and platonic solids were all intensively researched and valued by ancient philosophers prior to Euclid. From Leonardo da Vinci and M. C. Escher through Picasso and others, symmetry happens naturally and has been portrayed by artists throughout history. The relationship between symmetry and geometry sparked considerable debate in the second half of the 19th century. According to Felix Klein's Erlangen curriculum, symmetry, as represented by the concept of a transformation group, precisely describes geometry. In traditional Euclidean geometry, symmetries are often represented by fixed transformations and congruences. Collineations, which modify the orientation of a straight line in projective geometry, perform a similar function. Klein's concept to "define a geometry via its symmetry group" was influenced by the new geometries of Bolyai and Lobachevsky, Riemann, Clifford, and Klein, and Sophocles Lie. Topology and geometric group theory value discrete symmetry but Lie theory and Riemannian geometry value continuous symmetry (Ong, 2020).

In projective geometry, the duality principle is a novel kind of symmetry. This meta-phenomenon may be defined informally as follows: any theorem is still true if the terms "point" and "plane" and "join" and "meet" and "contain" and "lie" are changed. A vector space's duality and dual space are related and similar.

3.5. LINEAR ALGEBRA

Linear algebra is the basis for practically all other branches of mathematics. Linear algebra, for example, is often utilized in current presentations of geometry to accurately specify basic shapes such as lines, planes, and rotations. Furthermore, the mathematical area known as functional analysis may be thought of as linear algebra applied to function spaces. Functional examination. Linear algebra is widely used in STEM fields because it makes it simpler to mimic a wide range of natural events and enhances the efficiency with which these models can be implemented on computers. It is standard practice to employ the knowledge that the differential of a multivariate function at a point is the linear map that most nearly approximates the function locally when working with first-order approximations for nonlinear systems that cannot be defined by linear algebra. This enables the issue to be tackled in a more manageable way (Nathan, 2012).

3.6. ABSTRACT ALGEBRA

In the context of mathematics, the study of algebraic structures is known as abstract algebra or current algebra. Structures found in algebra include groups, rings, fields, modules, vector spaces, lattices, and algebras over a field. The phrase "abstract algebra," which refers to the process of substituting variables for numbers in computation and reasoning, was used in the early 20th century to distinguish this branch of algebra from older subfields of algebra such as "fundamental algebra." The phrase "abstract algebra" is currently more often used to refer to primary or secondary school mathematics courses, rather than those taught at the college or university level. The homomorphisms shown by mathematical categories and the algebraic structures they include are what distinguishes them as unique entities. The formalization of category theory allows for a consistent manner of expressing features and constructs shared by many different kinds of structures. This is made feasible since category theory has been formalized. The discipline of universal algebra, which is related to algebra, focuses on the study of algebraic structures of all kinds. In universal algebra, for example, the group structure or the variety of groups constitutes a separate object. This is also true for many other forms of algebra (Ong, 2020).

CHAPTER 4

Modular Arithmetic and Polynomials

Contents

4.1. INTRODUCTION

Modular arithmetic is a branch of number theory that deals with the arithmetic operations performed within a fixed modulus. In modular arithmetic, numbers "wrap around" a fixed range called the modulus, resulting in a finite set of possible remainders. This concept is often represented using the notation "a ≡ b (mod m)," which means that a and b leave the same remainder when divided by m (Prešern, 2011)

The key properties of modular arithmetic include congruence relations, modular addition, modular multiplication, and modular exponentiation. Congruence relations establish equivalence classes among numbers with respect to a given modulus. Modular addition involves adding two numbers within the modulus, while modular multiplication involves multiplying two numbers and taking the result modulo the given modulus. Modular exponentiation allows for efficient calculations of large powers modulo a given modulus.

Polynomials are algebraic expressions consisting of variables, coefficients, and exponents, combined using addition, subtraction, and multiplication operations. A polynomial is typically represented as a sum of terms, where each term consists of a coefficient multiplied by a variable raised to a non-negative integer exponent (Hsu, 2018).

Polynomials play a fundamental role in various areas of mathematics, including algebra, calculus, and number theory. They are used to model relationships between variables, approximate functions, solve equations, and represent geometric curves. Examples of polynomials include linear polynomials (degree 1), quadratic polynomials (degree 2), and cubic polynomials (degree 3).

The integration of modular arithmetic with polynomial arithmetic offers several benefits and applications. One of the key advantages is that modular arithmetic allows for efficient calculations and reductions of polynomial coefficients, particularly when dealing with large integers. By performing operations modulo a chosen modulus, the size of intermediate results can be significantly reduced, simplifying computations and storage requirements (Prešern, 2011).

Modular arithmetic also plays a crucial role in error detection and correction algorithms. For instance, in polynomial-based error-correcting codes such as Reed-Solomon codes, modular arithmetic is used to encode and decode messages, allowing for reliable data transmission even in the presence of errors.

Furthermore, modular arithmetic in polynomial arithmetic finds applications in cryptography. Polynomial-based cryptographic schemes, such as the RSA algorithm, rely on modular arithmetic operations to ensure secure communication, data encryption, and digital signatures. The use of modular arithmetic provides a mathematical foundation for cryptographic protocols and enhances the security of sensitive information (Hsu, 2018).

The combination of modular arithmetic and polynomials enables efficient computations, error detection and correction, and secure communication in various mathematical and practical applications. Understanding the principles and techniques of modular arithmetic and its integration with polynomial arithmetic is essential for unlocking the full potential of these fields in solving real-world problems.

4.2. MODULAR ARITHMETIC

Modular arithmetic is a mathematical system that deals with the arithmetic operations performed within a fixed modulus. It provides a framework for manipulating numbers by considering their remainders when divided by a given modulus. This section explores the definition, properties, and practical applications of modular arithmetic.

Modular arithmetic involves congruence relations, modular addition, modular multiplication, and modular exponentiation. Congruence relations establish equivalence classes among numbers with respect to a given modulus (Hsu, 2018).

4.2.1. Congruence Relations

Congruence is a fundamental concept in modular arithmetic. Two numbers, a and b, are said to be congruent modulo m (denoted as "$a \equiv b \pmod{m}$") if they have the same remainder when divided by m. In other words, a and b differ by a multiple of m. For example, $11 \equiv 2 \pmod{3}$ because both numbers leave a remainder of 2 when divided by 3 (Prešern, 2011). Formally, $a \equiv b \pmod{m}$ if and only if m divides $(a - b)$. This can be expressed as $(a - b) \bmod m = 0$.

Congruence relations exhibit several important properties:

- **Reflexive Property:** $a \equiv a \pmod{m}$.
- **Symmetric Property:** If $a \equiv b \pmod{m}$, then $b \equiv a \pmod{m}$.
- **Transitive Property:** If $a \equiv b \pmod{m}$ and $b \equiv c \pmod{m}$, then $a \equiv c \pmod{m}$.

- **Addition Property:** If a ≡ b (mod m) and c ≡ d (mod m), then a + c ≡ b + d (mod m).
- **Multiplication Property:** If a ≡ b (mod m) and c ≡ d (mod m), then a × c ≡ b × d (mod m).

These properties allow for manipulation and simplification of congruence equations.

1. **Modular Addition:** It involves adding two numbers within the modulus. Given a ≡ b (mod m) and c ≡ d (mod m), the modular sum is defined as (a + c) ≡ (b + d) (mod m). This property allows for adding numbers modulo m without considering their actual values, focusing only on their remainders.
2. **Modular Multiplication:** It is similar to modular addition but involves multiplying two numbers within the modulus. For example, if a ≡ b (mod m) and c ≡ d (mod m), then (a × c) ≡ (b × d) (mod m). Modular multiplication is particularly useful in reducing the size of intermediate results in calculations involving large numbers (Hsu, 2018).
3. **Modular Exponentiation:** It is the process of raising a number to a power modulo m. For example, given a ≡ b (mod m), the congruence (a^k) ≡ (b^k) (mod m) holds true for any positive integer k. Modular exponentiation is often employed in cryptography and number theory applications due to its efficiency and properties (Prešern, 2011).

4.2.2. Modular Arithmetic in Practice

Modular arithmetic finds practical applications in solving congruence equations, finding modular inverses, and employing the Chinese Remainder Theorem (CRT).

1. **Solving Congruence Equations:** It involves finding the values of variables that satisfy the given congruence relation. Congruence equations are typically expressed in the form of "a ≡ b (mod m)," where a, b, and m represent integers.

Several methods can be employed to solve congruence equations. One approach is trial and error, where potential values for the variables are tested systematically until a solution is found. This method can be effective for simple congruence equations with small modulus values.

Substitution is another method commonly used to solve congruence equations. By substituting known values for variables in the congruence relation, the equation can be simplified and the value of the remaining variable can be determined.

Algebraic manipulation is a more systematic approach to solving congruence equations. It involves applying mathematical operations such as addition, subtraction, multiplication, and division to both sides of the congruence relation in order to isolate the variable and find its value.

In more complex cases, advanced techniques such as modular inverses, the CRT, or Euler's theorem may be employed to find solutions to congruence equations. The choice of method depends on the specific congruence equation and its complexity. It is important to consider the properties of congruence relations and modular arithmetic to ensure accurate and valid solutions.

2. **Modular Inverses:** These are fundamental in modular arithmetic and have significant applications. Given a number a modulo m, the modular inverse of a, denoted as a^(–1), is another number x such that (a × x) ≡ 1 (mod m). In other words, the product of and its modular inverse is congruent to 1 modulo m.

Modular inverses are employed in various scenarios. First, they are utilized in solving congruence equations. When a congruence equation involves division modulo m, finding the modular inverse allows for the division operation to be replaced with multiplication. This simplifies the equation and enables the determination of the variable's value (Hsu, 2018).

Additionally, modular inverses are crucial in certain cryptographic algorithms. In asymmetric encryption schemes like RSA, modular inverses play a central role in generating and manipulating keys. The modular inverse of a number is used to encrypt or decrypt data, ensuring the security and confidentiality of the communication (Prešern, 2011).

Moreover, modular inverses find applications in modular exponentiation algorithms. When calculating large powers modulo m, modular inverses can be employed to reduce the number of modular multiplications required, leading to more efficient computations.

The computation of modular inverses can be achieved using various methods, such as the extended Euclidean algorithm or Fermat's little theorem, depending on the specific modulus and number involved.

3. **Chinese Remainder Theorem (CRT):** It is a versatile and powerful tool in modular arithmetic. It provides a solution to a

system of simultaneous congruences, where each congruence is defined with respect to a different modulus. The theorem enables the reconstruction of a number's value from its remainders modulo different moduli.

The CRT has applications in various fields, including cryptography, data encryption, and solving mathematical puzzles. In cryptography, it is utilized in public-key encryption algorithms such as RSA. The CRT allows for efficient decryption by breaking down the encryption process into smaller calculations modulo different prime moduli (Prešern, 2011).

In data encryption, the CRT plays a vital role in the CRT-based encryption schemes. These schemes provide a more efficient and secure way to encrypt and decrypt data, especially in scenarios where the moduli are large and computationally intensive. Moreover, the CRT is employed in solving mathematical puzzles and problems. It enables finding solutions that satisfy multiple congruences simultaneously, allowing for efficient and accurate computations. The CRT establishes a direct connection between the remainders of a number modulo different moduli and the number itself. It provides a constructive method to reconstruct the original number using the CRT algorithm, which involves solving a system of congruences and combining the solutions using modular arithmetic operations. Modular arithmetic involves congruence relations, modular addition, modular multiplication, and modular exponentiation. It offers a systematic approach to perform arithmetic operations within a fixed modulus, enabling efficient calculations and simplification of results. Practical applications of modular arithmetic include solving congruence equations, finding modular inverses, and employing the CRT in various fields, including cryptography and number theory (Hsu, 2018).

4.3. POLYNOMIALS

Polynomials are essential mathematical objects that play a crucial role in various fields, including algebra, calculus, and number theory. This section explores the basic concepts of polynomials, polynomial operations, polynomial factorization, and roots of polynomials.

4.3.1. Basic Concepts of Polynomials

A polynomial is an algebraic expression consisting of variables, coefficients, and exponents, combined using addition, subtraction, and multiplication operations. It is typically written as a sum of terms, where each term consists of a coefficient multiplied by a variable raised to a non-negative integer exponent.

For example, the polynomial $f(x) = 3x^2 + 2x - 1$ consists of three terms: $3x^2$, $2x$, and -1. The variable x is raised to the powers of 2, 1, and 0, respectively, and the coefficients are 3, 2, and –1.

4.3.1.1. Polynomial Degree

The degree of a polynomial is a significant characteristic that influences several aspects of polynomial arithmetic and analysis. It represents the highest power of the variable in any of the polynomial's terms and provides insight into the "size" or "complexity" of the polynomial.

For instance, consider the polynomial $f(x) = 3x^2 + 2x - 1$. In this case, the highest power of x is 2, which means that the polynomial has a degree of 2. The degree helps classify the polynomial as a quadratic polynomial.

The degree of a polynomial has implications for various aspects of polynomial arithmetic and analysis. It determines the number of possible roots of the polynomial. In general, a polynomial of degree n can have at most n roots, including real and complex roots. For example, a quadratic polynomial can have up to two roots (Prešern, 2011). Additionally, the degree of a polynomial influences the behavior of the polynomial function. Higher-degree polynomials exhibit more complex behavior, such as multiple turning points, varying concavity, and a larger number of intersections with the x-axis.

The degree also plays a role in polynomial long division and synthetic division. These methods are used to simplify and factorize polynomials, and the degree helps determine the number of steps required for the division process.

Moreover, the degree affects the rate at which the polynomial grows or decays as the variable's magnitude increases. It influences the end behavior of the polynomial function, determining whether the function approaches positive or negative infinity as the variable approaches positive or negative infinity (Hsu, 2018).

4.3.1.2. Polynomial Coefficients

Polynomial coefficients are essential components of a polynomial and determine the specific values that the polynomial takes on when the variable is substituted with particular values. In the polynomial f(x) = 3x^2 + 2x – 1, the coefficients are 3, 2, and –1 (Prešern, 2011).

Polynomial coefficients can be any real or complex numbers. They provide the scaling factors that determine the magnitude and contribution of each term in the polynomial expression. By multiplying the variable raised to a specific exponent by its coefficient, the polynomial coefficients influence the shape, position, and behavior of the polynomial curve.

The coefficients can affect various properties of the polynomial, such as the location of the polynomial's zeros (roots), the shape of the graph, the symmetry of the polynomial, and the relationship between different terms.

In addition to their influence on the polynomial's properties, the coefficients play a significant role in polynomial operations. When performing operations such as addition, subtraction, multiplication, and division of polynomials, the coefficients of corresponding terms are combined and manipulated accordingly.

Moreover, the coefficients are crucial in polynomial factorization. The factorization process involves expressing a polynomial as a product of simpler factors, which allows for a deeper understanding of the polynomial's behavior and facilitates solving polynomial equations (Hsu, 2018).

The choice of coefficients can lead to polynomials with different characteristics and behaviors. Adjusting the coefficients can shift the position of the polynomial curve, change the number and nature of its roots, or affect its symmetry.

4.3.2. Polynomial Operations

Polynomial operations involve manipulating polynomials using addition, subtraction, and multiplication.

4.3.2.1. Polynomial Addition and Subtraction

Polynomial addition involves adding the corresponding terms of two polynomials. For example, to add the polynomials f(x) = 3x^2 + 2x – 1 and g(x) = 2x^2 – x + 4, we combine the like terms to obtain the sum h(x) = 5x^2 + x + 3.

Polynomial subtraction is similar to polynomial addition, but with the subtraction of corresponding terms. For example, to subtract g(x) from f(x), we subtract the coefficients of like terms to obtain the difference h(x) = x^2 + 3x – 5 (Prešern, 2011).

4.3.2.2. Polynomial Multiplication

Polynomial multiplication involves multiplying each term of one polynomial by each term of the other polynomial and then combining like terms. For example, to multiply the polynomials f(x) = 3x^2 + 2x – 1 and g(x) = 2x^2 – x + 4, we perform the multiplication and combine like terms to obtain the product h(x) = 6x^4 + 5x^3 – 4x^2 + 11x – 4. Polynomial multiplication is essential for various applications, such as finding areas and volumes, solving equations, and approximating functions.

4.3.2.3. Polynomial Division

Polynomial division involves dividing one polynomial by another polynomial to obtain a quotient and a remainder. The process is analogous to long division. For example, to divide the polynomial f(x) = 6x^3 + 3x^2 – 2x + 1 by the polynomial g(x) = 2x – 1, we perform polynomial long division to obtain the quotient h(x) = 3x^2 + 4x + 2 and a remainder of 3 (Hsu, 2018). Polynomial division allows for simplifying complex polynomial expressions and solving equations involving polynomials.

4.3.3. Polynomial Factorization

Polynomial factorization involves expressing a polynomial as a product of simpler polynomials or irreducible factors. It is a crucial technique for simplifying polynomial expressions, solving equations, and understanding the behavior of polynomial functions.

4.3.3.1. Factor Theorem

The factor theorem states that if a polynomial f(x) has a root c, then (x – c) is a factor of f(x). This theorem provides a useful method for identifying factors and simplifying polynomial expressions.

4.3.3.2. Polynomial Long Division

Polynomial long division is a process used to factorize polynomials. It involves dividing the polynomial by a linear factor repeatedly until the polynomial is fully factorized or a remainder of zero is obtained.

4.3.3.3. Synthetic Division

Synthetic division is a streamlined method for polynomial division when dividing by a linear factor of the form (x – c). It allows for more efficient calculations and simplifies the process of finding factors and determining remainders.

4.3.3.4. Quadratic Polynomials

Quadratic polynomials are polynomials of degree 2. They play a significant role in algebra, calculus, and geometry. Quadratic polynomials have various properties and can be factored, graphed, and used to solve quadratic equations (Prešern, 2011).

4.3.4. Roots of Polynomials

4.3.4.1. Finding Roots

This is a fundamental task in algebra, calculus, and numerical analysis. The roots of a polynomial are the values of the variable that make the polynomial equal to zero. These values have significant implications for solving equations, analyzing the behavior of the polynomial function, and understanding its graphical representation (Hsu, 2018). The process of finding polynomial roots can be approached using various root-finding techniques. Some common methods include factoring, synthetic division, and numerical methods like Newton's method. Factoring is a straightforward technique that can be used when the polynomial can be easily factored. By factoring the polynomial into linear or quadratic factors, the roots can be identified directly. Synthetic division is particularly useful when searching for rational roots of a polynomial. This method allows for efficient evaluation of the polynomial at a given value, making it possible to test potential roots quickly and narrow down the possibilities. In cases where factoring and synthetic division are not feasible, numerical methods such as Newton's method or the bisection method can be employed. These techniques involve iterative calculations to approximate the roots of the polynomial. They provide numerical solutions and are applicable to polynomials of any degree.

Furthermore, advanced algorithms and software packages are available for finding roots of high-degree polynomials or cases where the roots cannot be expressed analytically.

4.3.4.2. Rational and Irrational Roots

The roots of polynomials can be categorized as either rational or irrational numbers. Rational roots are those that can be expressed as a ratio of two integers, while irrational roots cannot be expressed in such a form. Understanding the distinction between rational and irrational roots is crucial in analyzing polynomial equations (Prešern, 2011).

The Rational Root Theorem is a useful tool for identifying potential rational roots of a polynomial equation. According to this theorem, if a polynomial equation with integer coefficients has a rational root p/q, where p and q are integers, then p must be a factor of the constant term and q must be a factor of the leading coefficient. By testing the factors of the constant term and the leading coefficient, potential rational roots can be identified.

However, it is important to note that not all polynomials have rational roots. Many polynomials, especially those of higher degree, have one or more irrational roots that cannot be expressed as a ratio of two integers. These roots often involve numbers such as square roots, cube roots, or other non-terminating decimal expansions (Hsu, 2018).Rational roots are particularly valuable in polynomial equations as they can provide insight into the integer solutions of the equation. By finding rational roots, it becomes possible to simplify the equation and determine potential values for the variable.

On the other hand, irrational roots introduce more complexity and often require numerical methods or approximation techniques to estimate their values. These roots can have profound implications for the behavior of the polynomial function, such as the presence of local extrema or points of inflection.

4.3.4.3. Complex Roots

Polynomials can also have complex roots, which involve imaginary numbers. Complex roots occur in conjugate pairs, where if one root is $a + bi$, the other root is $a - bi$, where a and b are real numbers and i is the imaginary unit ($\sqrt{(-1)}$).

Understanding the properties and methods of finding roots allows for solving equations, analyzing polynomial functions, and using polynomials to model real-world phenomena.

4.4. MODULAR ARITHMETIC WITH POLYNOMIALS

Modular arithmetic with polynomials combines the principles of modular arithmetic and polynomial arithmetic. This section delves into the introduction to modular polynomials, polynomial arithmetic modulo n, modular polynomial operations, and applications of modular polynomial arithmetic. Modular polynomials are polynomials in which all arithmetic operations are performed modulo a given integer, referred to as the modulus. They are constructed by taking the coefficients of the polynomial and reducing them modulo the given modulus. The resulting polynomial follows the rules of modular arithmetic and has coefficients within a restricted range (Prešern, 2011). Modular polynomials find applications in various areas, such as cryptography, error detection and correction, and computer science. By performing arithmetic operations modulo a chosen modulus, modular polynomials allow for efficient computations and storage of large polynomial coefficients.

4.4.1. Polynomial Arithmetic Modulo n

Polynomial arithmetic modulo n involves performing arithmetic operations on polynomials, where each coefficient is reduced modulo n. In this context, n represents the modulus or the chosen integer. When adding or subtracting polynomials modulo n, the corresponding coefficients are added or subtracted modulo n, respectively. For example, in polynomial addition modulo 7, $(3x^2 + 2x - 1) + (2x^2 - x + 4) = 5x^2 + x + 3$ (mod 7) (Hsu, 2018). Similarly, polynomial multiplication modulo n is performed by multiplying the coefficients modulo n and reducing the result modulo n. For instance, in polynomial multiplication modulo 5, $(3x^2 + 2x - 1) \times (2x^2 - x + 4) = x^4 + 4x^3 + 2x^2 + 3$ (mod 5).

4.4.2. Modular Polynomial Operations

4.4.2.1. Modular Polynomial Addition

Modular polynomial addition involves adding the corresponding coefficients of two polynomials modulo the chosen modulus. This operation follows the principles of modular arithmetic. For example, $(3x^2 + 2x - 1) + (2x^2 - x + 4) = 5x^2 + x + 3$ (mod n).

Modular polynomial addition allows for efficient computation and storage of polynomial coefficients by reducing them modulo the modulus (Prešern, 2011).

4.4.2.2. Modular Polynomial Multiplication

Modular polynomial multiplication is performed by multiplying the coefficients of two polynomials modulo the chosen modulus. The resulting polynomial has coefficients reduced modulo the modulus. For example, (3x^2 + 2x – 1) × (2x^2 – x + 4) = x^4 + 4x^3 + 2x^2 + 3 (mod n).

Modular polynomial multiplication is crucial in various applications, such as error correction codes and cryptographic algorithms, as it allows for efficient calculations with reduced coefficient sizes.

4.4.2.3. Modular Polynomial Exponentiation

Modular polynomial exponentiation involves raising a polynomial to a power modulo the chosen modulus. The exponentiation operation follows the principles of modular arithmetic. For example, (3x^2 + 2x – 1)^3 = 27x^6 + 54x^5 + 36x^4 + 12x^3 – 9x^2 – 6x + 1 (mod n). Modular polynomial exponentiation finds applications in various areas, including cryptography, where it enables efficient encryption and decryption operations (Hsu, 2018).

4.4.3. Applications of Modular Polynomial Arithmetic

4.4.3.1. Cryptography

Cryptography is a field that focuses on secure communication and data protection. Modular polynomial arithmetic is integral to various cryptographic schemes, providing the foundation for encryption and decryption algorithms such as the RSA algorithm.

In cryptography, modular polynomial arithmetic ensures secure communication by performing arithmetic operations on polynomials modulo a chosen modulus. This modular reduction limits the size of coefficients and allows for efficient computations while preserving the underlying mathematical structure.

The RSA algorithm, one of the most widely used public-key encryption schemes, relies on modular polynomial arithmetic. The algorithm utilizes modular exponentiation, where a polynomial is raised to a power modulo a chosen modulus. This operation is computationally efficient and forms the basis for encryption and decryption processes (Prešern, 2011).

In the RSA encryption process, a message is encoded as a polynomial, and the recipient's public key, consisting of a modulus and an encryption exponent, is used to encrypt the polynomial. Modular polynomial multiplication is employed to perform the encryption, reducing the coefficients modulo the modulus. On the recipient's end, modular polynomial arithmetic is again utilized in the RSA decryption process. The recipient's private key, comprising a modulus and a decryption exponent, is applied to the encrypted polynomial, using modular polynomial exponentiation to recover the original message.

The security of cryptographic schemes, such as RSA, relies on the difficulty of factoring large numbers. Modular polynomial arithmetic ensures that the operations involved in encryption and decryption are efficient and can be performed within reasonable time frames, while preserving the confidentiality and integrity of the encrypted data.

4.4.3.2. Error Detection and Correction

Error detection and correction techniques are crucial for ensuring reliable data transmission, especially in scenarios where errors are likely to occur during communication. Modular polynomial arithmetic plays a significant role in error detection and correction codes, such as Reed-Solomon codes.

Reed-Solomon codes are widely used for error detection and correction in various applications, including data storage systems, wireless communication, and satellite communication. These codes utilize modular polynomials to encode and decode messages, enabling the detection and correction of errors introduced during transmission (Hsu, 2018).

Modular polynomial arithmetic is employed in Reed-Solomon codes to encode the original message into a polynomial form. The coefficients of the polynomial are determined by evaluating the message at specific points, often referred to as evaluation points. The evaluations are performed using modular arithmetic, reducing the coefficients modulo a chosen modulus.

When errors occur during transmission, modular polynomial arithmetic is used to detect and correct these errors. The received polynomial is evaluated at the same set of evaluation points, resulting in a set of received values. By comparing the received values with the expected values based on the original polynomial, errors can be identified.

To correct the errors, polynomial interpolation techniques are employed. These techniques involve finding a polynomial that fits the received values as closely as possible. The corrected polynomial is then used to recover the original message (Prešern, 2011).

Modular polynomial arithmetic ensures the integrity and accuracy of the data by providing a framework for error detection and correction. The modular reduction of polynomial coefficients allows for efficient computations and compact representation of encoded messages.

4.4.3.3. Polynomials in Computer Science

From computer graphics and data compression to coding theory, polynomials provide a versatile mathematical tool for representing curves, analyzing data structures, and developing efficient algorithms.

In computer graphics, polynomials are extensively used to represent curves and surfaces. Techniques like Bézier curves and splines utilize polynomials to define smooth and precise shapes in computer-generated images. Modular polynomial arithmetic allows for efficient manipulation and evaluation of these curves, enabling realistic and visually appealing graphics (Hsu, 2018).

Data compression techniques, such as polynomial interpolation and regression, leverage polynomials to approximate and compress large datasets. By representing the data with a polynomial, it becomes possible to reduce its size while preserving essential characteristics. Modular polynomial arithmetic ensures efficient compression and decompression processes, allowing for the storage and transmission of data in a more compact form.

Coding theory, which deals with error detection and correction codes, heavily relies on polynomials.

Linear block codes, such as Reed-Solomon codes, employ modular polynomial arithmetic to encode and decode messages, providing robust error correction capabilities. Polynomials are used to represent code words, error patterns, and check polynomials, enabling efficient error detection and correction algorithms.

Furthermore, polynomials find applications in polynomial-time algorithms, computational complexity theory, and algorithmic design. Polynomials are employed to analyze the running time and complexity of algorithms, providing insights into their efficiency and scalability. Modular polynomial arithmetic is utilized to develop efficient algorithms for solving computational problems, such as polynomial interpolation, polynomial factorization, and polynomial root finding.

4.5. POLYNOMIALS AND FINITE FIELDS

Polynomials and finite fields are intimately connected; as finite fields provide the mathematical framework for polynomial arithmetic in a restricted set of elements. This section explores the introduction to finite fields, their properties, construction, and the application of polynomial arithmetic within finite fields.

A finite field, also known as a Galois field, is a mathematical structure consisting of a finite set of elements along with two operations: addition and multiplication. The elements in a finite field form a field, meaning that they satisfy specific properties, such as closure, associativity, commutativity, and the existence of inverses (Prešern, 2011).

Finite fields have a finite number of elements, denoted by q, where q is a prime power or a power of a prime. The order of a finite field is the number of elements it contains. One fundamental property of finite fields is that the characteristic of the field, denoted by char(F), is a prime number. The characteristic represents the number of times one must add the multiplicative identity (typically denoted as 1) to itself to obtain the additive identity (typically denoted as 0).

4.5.1. Finite Field Construction

Finite fields can be constructed using different methods, such as the primitive polynomial method, the irreducible polynomial method, or the polynomial quotient method (Hsu, 2018). The primitive polynomial method constructs a finite field by taking a primitive polynomial of degree m over a base field (typically the field of integers modulo a prime) and considering the field extension generated by the roots of this polynomial.

The irreducible polynomial method constructs a finite field by taking an irreducible polynomial of degree m over a base field and considering the field extension generated by the coefficients of the polynomial modulo the base field.

The polynomial quotient method constructs a finite field by taking a polynomial of degree m over a base field and considering the field extension generated by the equivalence classes of polynomials modulo a given irreducible polynomial.

These construction methods ensure that the resulting finite field is well-defined, satisfies the properties of a field, and contains q elements, where q is a prime power or a power of a prime.

4.5.2. Polynomial Arithmetic in Finite Fields

4.5.2.1. Addition, Subtraction, and Multiplication

Polynomial arithmetic in finite fields follows the principles of modular arithmetic. Addition and multiplication of polynomials in a finite field involve performing the operations on the coefficients modulo the characteristic of the field (Prešern, 2011).

In finite fields, polynomial addition and subtraction are performed by adding or subtracting the corresponding coefficients of the polynomials modulo the characteristic of the field. For example, in a finite field with characteristic 5, $(3x^2 + 2x - 1) + (2x^2 - x + 4) = 0x^2 + 1x - 2 \pmod 5$.

Multiplication of polynomials in a finite field is performed by multiplying the coefficients modulo the characteristic and reducing the result modulo an irreducible polynomial that defines the finite field. For example, in a finite field with characteristic 7 and an irreducible polynomial $g(x) = x^3 + 2x + 1$, $(3x^2 + 2x - 1) \times (2x^2 - x + 4) = 3x^4 + 4x^3 + 6x^2 + 5x + 6 \pmod{g(x)}$.

4.5.2.2. Polynomial Division in Finite Fields

Polynomial division in finite fields is a crucial operation that allows for the determination of the quotient and remainder when dividing one polynomial by another within the context of a finite field. The process bears similarities to polynomial long division, but with some key distinctions (Hsu, 2018).

In finite fields, the coefficients involved in polynomial division are reduced modulo the characteristic of the field. Additionally, the divisor is an irreducible polynomial of the finite field. These considerations ensure that the division process remains within the finite field and adheres to the properties and constraints of modular arithmetic.

The division process begins by multiplying the divisor by an appropriate polynomial, known as the quotient polynomial. This product is then subtracted from the dividend, resulting in a new polynomial. At each step, the coefficients of the polynomials involved are reduced modulo the characteristic of the finite field.

The process of multiplying the divisor by the quotient polynomial and subtracting it from the dividend is repeated until the degree of the remainder polynomial becomes smaller than the degree of the divisor polynomial. The final remainder obtained represents the residue of the division.

Polynomial division in finite fields is essential in various applications, such as error correction codes and cryptography. It enables the computation of error syndromes and the recovery of original messages from received polynomial representations. Additionally, it plays a role in polynomial factorization and determining the roots of polynomials over finite fields.

4.5.2.3. Polynomial Factorization in Finite Fields

Polynomial factorization in finite fields is a powerful tool with applications in error correction codes, cryptography, and number theory. Factorization involves expressing a polynomial as a product of irreducible polynomials over the finite field (Prešern, 2011).

In finite fields, due to the limited number of elements, every polynomial can be factored into irreducible polynomials. The process of polynomial factorization in finite fields often employs techniques like the Berlekamp's algorithm or the Cantor-Zassenhaus algorithm.

These algorithms iteratively find factors by determining the roots of the polynomial and reducing it modulo the found factors. This process is repeated until the polynomial is fully factored into irreducible components.

Polynomial factorization in finite fields has numerous applications, including constructing error correction codes, generating cryptographic keys, and solving mathematical problems related to number theory (Hsu, 2018).Polynomials and finite fields are closely intertwined, with finite fields providing the framework for polynomial arithmetic within a restricted set of elements. Finite fields possess unique properties and can be constructed using different methods. Polynomial arithmetic in finite fields involves addition, subtraction, multiplication, division, and factorization, with coefficients reduced modulo the characteristic and irreducible polynomials defining the field. The application of polynomial arithmetic in finite fields finds utility in various fields, including error correction codes, cryptography, and number theory.

4.6. POLYNOMIALS AND MODULAR EQUATIONS

Polynomials and modular equations are interconnected topics that involve solving polynomial equations, working with polynomial congruences, and solving systems of polynomial equations. This section explores the concepts of solving polynomial equations modulo n, polynomial congruences, and the application of polynomial techniques in solving systems of polynomial equations, including the CRT, polynomial interpolation, and cryptography.

4.6.1. Solving Polynomial Equations Modulo n

Solving polynomial equations modulo n involves finding solutions to equations where the variables and coefficients are integers reduced modulo n. This approach allows for efficient computations and simplification of equations by restricting the solutions to a finite set of remainders (Hsu, 2018).

To solve a polynomial equation modulo n, one can employ various techniques, including trial and error, substitution, and algebraic manipulation. These methods involve systematically testing different values or applying algebraic transformations to simplify the equation and identify solutions that satisfy the congruence relation modulo n.

The solutions to polynomial equations modulo n are typically expressed as congruence classes, representing the set of all integers that are congruent to the solution modulo n. These congruence classes can be represented as residue classes, often denoted by square brackets, such as [x]_n, where x represents the solution and n is the modulus (Prešern, 2011). Solving polynomial equations modulo n finds applications in cryptography, number theory, and computer science, providing a foundation for secure communication, primality testing, and algorithm design.

4.6.2. Polynomial Congruences

Polynomial congruences involve the study of congruence relations between polynomials. A polynomial congruence states that two polynomials are congruent modulo n if their coefficients are congruent modulo n.

The concept of polynomial congruences builds upon the principles of modular arithmetic, extending them to polynomial expressions. It allows for the manipulation of polynomials within a modular framework, reducing coefficients and performing operations modulo a given modulus.

Polynomial congruences find applications in solving polynomial equations modulo n, analyzing polynomial properties, and constructing error correction codes. They provide a systematic approach to working with polynomials within a modular context, facilitating computations and simplifying polynomial operations.

4.6.3. Solving Systems of Polynomial Equations

Solving systems of polynomial equations involves finding solutions that satisfy multiple polynomial equations simultaneously. This topic is of great significance in various fields, including mathematics, engineering, and computer science, as many real-world problems can be formulated as systems of polynomial equations. Several techniques can be employed to solve systems of polynomial equations, such as the CRT, polynomial interpolation, and specialized algorithms designed for specific problem domains (Hsu, 2018).

4.6.3.1. Chinese Remainder Theorem (CRT) and Polynomials

The CRT is a powerful tool in solving systems of polynomial equations. It provides a method for finding solutions to a system of simultaneous congruences, allowing for the reconstruction of a number's value from its remainders modulo different moduli.

When applied to polynomial equations, the CRT enables the solution of systems of polynomial congruences, where the coefficients of the polynomials are reduced modulo different moduli. By solving each congruence individually and applying the CRT, the solutions to the system of polynomial equations can be determined. The CRT has diverse applications in cryptography, number theory, and algorithm design. It allows for efficient computations in modular arithmetic and provides a framework for solving complex systems of polynomial equations with multiple variables (Prešern, 2011).

4.6.3.2. Polynomial Interpolation

Polynomial interpolation is another method used to solve systems of polynomial equations. It involves finding a polynomial that passes through a given set of data points, determining the coefficients of the polynomial that satisfies the equations. Polynomial interpolation techniques, such as Lagrange interpolation and Newton interpolation, provide algorithms for constructing the interpolating polynomial based on the known data points. These methods utilize the values of the data points and their corresponding variables to determine the polynomial coefficients. By employing polynomial interpolation, systems of polynomial equations can be transformed into a problem of finding an interpolating polynomial, thereby enabling the determination of the unknown variables. Polynomial interpolation has diverse applications in various fields, including numerical analysis, computer graphics, data fitting, and signals processing. It allows for the estimation and prediction of missing data points, curve generation, and approximation of complex functions (Hsu, 2018).

4.6.3.3. Application in Cryptography

The techniques of solving systems of polynomial equations find valuable applications in cryptography, particularly in the design and analysis of cryptographic algorithms. Polynomial equations play a crucial role in modern cryptographic schemes, including those used in encryption, decryption, digital signatures, and secure communication.

In cryptography, systems of polynomial equations are often solved using methods such as the CRT and polynomial interpolation. These techniques enable the efficient computation of cryptographic keys, the secure generation of random numbers, and the implementation of encryption and decryption algorithms (Prešern, 2011).

The application of polynomial techniques in cryptography involves solving systems of polynomial equations modulo prime numbers or composite numbers, using modular arithmetic and congruence relations to find solutions that ensure the security and integrity of the cryptographic protocols.

Polynomials and modular equations are intertwined topics that encompass solving polynomial equations modulo n, working with polynomial congruences, and solving systems of polynomial equations. These concepts find applications in cryptography, number theory, computer science, and various problem domains. Techniques such as the CRT, polynomial interpolation, and modular arithmetic provide powerful tools for solving systems of polynomial equations, enabling efficient computations, secure communication, and advanced algorithm design (Hsu, 2018).

4.7. SUMMARY OF KEY POINTS

In this chapter, we explored the concept of modular arithmetic and its connection to polynomials. We delved into various subtopics, including an introduction to modular arithmetic and polynomials, the definition and properties of modular arithmetic, modular addition, multiplication, and exponentiation, solving congruence equations, modular inverses, and the CRT. We also discussed polynomials in depth, covering their basic concepts such as definition, degree, and coefficients. We explored polynomial operations like addition, subtraction, multiplication, and division, as well as polynomial factorization and finding roots. Additionally, we examined the relationship between polynomials and finite fields, as well as their applications in cryptography, error detection and correction, and computer science. Modular arithmetic and polynomials play vital roles in various domains. Modular arithmetic provides a powerful tool for computations with remainders, allowing for efficient calculations and simplified equations. It finds applications in cryptography, error detection and correction, computer science, and number theory. Polynomials, on the other hand, are versatile mathematical objects that are used to represent curves, analyze data structures, and develop efficient algorithms. They are fundamental in areas such as computer graphics, data compression, coding theory, and polynomial-time algorithms. The combination of modular arithmetic and polynomials, as seen in modular polynomial arithmetic and solving polynomial equations modulo n, extends the applicability of both concepts. It enables efficient computations, simplification of equations, error correction, secure communication, and the development of advanced algorithms (Prešern, 2011). The study of modular arithmetic and polynomials continues to evolve, leading to exciting future directions and applications. Some potential areas of exploration include:

1. **Cryptanalysis:** Further research can be done to explore the security of cryptographic schemes based on modular arithmetic and polynomials. This includes developing more efficient attacks and countermeasures to ensure the robustness of cryptographic systems.
2. **Algorithmic Development:** The development of advanced algorithms for polynomial operations, modular arithmetic, and solving systems of polynomial equations can enhance computational efficiency and enable the solution of complex problems in various fields (Hsu, 2018).
3. **Error Correction Codes:** Investigating new coding techniques and error correction algorithms based on modular polynomial arithmetic can lead to more reliable and efficient error detection and correction schemes for data storage and communication systems.
4. **Applications in Artificial Intelligence:** Exploring the integration of modular arithmetic and polynomials in machine learning algorithms, pattern recognition, and data analysis can provide new insights and tools for solving complex AI problems.

Modular arithmetic and polynomials are powerful mathematical tools with numerous applications. They enable efficient computations, secure communication, error detection and correction, and algorithm design. The future holds exciting possibilities for further research and development in these areas, leading to advancements in cryptography, computer science, data analysis, and artificial intelligence (Prešern, 2011).

CHAPTER 5

Mathematical Functions

Contents

5.1. INTRODUCTION

Mathematical functions are fundamental mathematical concepts that establish relationships between different data sets.

They play a pivotal role in various fields, providing a structured way to represent and analyze real-world phenomena. This section introduces the core ideas behind functions and their significance in diverse disciplines (Luke, 2014)

At its core, a function is a rule or mapping that associates each element from one set (the domain) to a unique component of another group (the codomain). The function is denoted by a symbol, such as f, and expressed as f: A → B, where A is the domain, and B is the codomain. The fundamental property of a function is that for each element a in A, only one corresponding element b exists in B.

1. **Function Notation:** Functions are typically represented using function notation, such as f(x) or y = f(x), where x is the independent variable, and y is the dependent variable. For example, the function f(x) = 2x + 3 represents a linear relationship between x and y, where x's value determines y's value (Saxena, 2011).
2. **Domain and Codomain:** The profession of a function is the set of all possible input values (independent variables) for which the process is defined. The codomain is the set of all possible output values (dependent variables) that the function can produce. It is essential to distinguish between the codomain and the range, where the content is the actual set of output values produced by the process.
3. **One-to-One and Onto Functions:** A function is said to be one-to-one (injective) if each element in the domain corresponds to a unique component of the codomain. In other words, no two different aspects in the field can map to the same part in the codomain. An onto function (surjective) is one where every detail in the codomain has at least one pre-image in the domain. In other words, the process covers the entire codomain.
4. **Inverse Functions:** An inverse function is a unique function that "undoes" the original function. If f(x) is a function, its inverse function, denoted as f^(–1)(x), performs the reverse operation. In other words, if f(a) = b, then f^(–1)(b) = a. Not all functions

have inverse functions, and the existence of an inverse function depends on the properties of the original function.

Functions establish a clear relationship between independent and dependent variables. The independent variable, usually denoted as x, represents the input value from the domain. In contrast, the dependent variable, typically denoted as y, means the output value in the codomain, which depends on the independent variable.

5. **Understanding Independent and Dependent Variables:** In a function, the independent variable is the variable that is chosen or controlled by the experimenter or the system. It is the "input" into the process. On the other hand, the dependent variable is the variable that changes in response to the independent variable. It is the "output" of the function (Luke, 2014).
6. **Real-World Examples of Functions:** Functions are prevalent in real-world applications. For example, in physics, the displacement of an object as a function of time is a classic example of a process. The distance traveled by car in a given time, the temperature change over time, and the growth rate of a population are all examples of functions.

Functions are fundamental tools in various scientific and engineering disciplines. They provide a systematic way to model and analyze relationships between variables, enabling accurate predictions and problem-solving (Saxena, 2011).

7. **Functions in Physics and Natural Sciences:** Physics relies heavily on parts to describe and predict physical phenomena. Roles are used to model the motion of objects, the behavior of waves, the distribution of particles, and the changes in energy and temperature over time. Functions play a critical role in formulating scientific laws and principles.
8. **Functions in Engineering:** Engineering heavily depends on functions to design and analyze systems. Functions describe the behavior of electrical circuits, mechanical systems, control processes, and fluid dynamics. Engineers use functions to optimize system performance, ensure stability, and predict system responses.
9. **Functions in Economics and Finance:** In economics, functions are used to model supply and demand curves, production functions, utility functions, and cost functions. These functions

help economists understand market behavior, predict economic trends, and make informed decisions. In finance, parts are used to model investment growth, risk assessment, and financial forecasting.

10. **Functions in Computer Science and Data Analysis:** In computer science, functions are central to algorithms and programming. They enable data transformation, data analysis, and decision-making processes. Data analysis functions are used for statistical modeling, machine learning, and data visualization.
11. **Functions in Geometry and Trigonometry:** In geometry, functions study geometric shapes and figures. Trigonometric functions, such as sine, cosine, and tangent, are crucial for calculating distances, angles, and rotations in 2D and 3D spaces. Roles are significant in describing the relationships between angles and sides in various geometric forms (Luke, 2014).

Introducing mathematical functions provides the foundational understanding of these essential concepts. Functions are potent tools in various fields, allowing us to model, analyze, and predict real-world phenomena. Their significance in physics, engineering, economics, computer science, and geometry showcases their versatility and importance in modern scientific and technological advancements.

5.2. FUNDAMENTALS OF FUNCTIONS

5.2.1. Types of Mathematical Functions

1. **Polynomial Functions:** These are functions composed of power functions with non-negative integer exponents. They are expressed as f(x) = a_nx^n + a_{n–1}x^{n–1} +... + a_1x + a_0, where a_n to a_0 are constants, and n is a positive integer. Polynomial functions can be of various degrees, such as linear (degree 1), quadratic (degree 2), cubic (degree 3), and so on. These functions are essential in algebra and calculus due to their simplicity and widespread applicability (Saxena, 2011).
2. **Trigonometric Functions:** These are functions derived from the ratios of sides of right triangles. The main trigonometric functions are sine (sin), cosine (cos), tangent (tan), cosecant (CSC), secant (sec), and cotangent (cot). These functions relate angles to the

lengths of sides in a right triangle and have periodic properties. Trigonometric functions are extensively used in geometry, physics, and engineering to describe rare phenomena such as oscillations and waves.

3. **Exponential Functions:** These are f(x) = a^x, where a is a positive constant (base), and x is the exponent. These functions have a unique property: the function's rate of change is proportional to its current value. Exponential functions find applications in growth and decay processes, compound interest calculations, and natural phenomena such as population growth.
4. **Logarithmic Functions:** These are the inverse of exponential functions. They are f(x) = log_a(x), where a is the base and x is the argument. The logarithm of a number x with floor a is the exponent a must be raised to obtain x. Logarithmic functions are crucial for solving exponential equations, analyzing data with varying scales, and in applications where quantities change by orders of magnitude (Luke, 2014).
5. **Hyperbolic Functions:** These are analogs of trigonometric functions defined using the hyperbola instead of the unit circle. The primary hyperbolic functions are sinh (hyperbolic sine), cosh (hyperbolic cosine), tanh (hyperbolic tangent), sech (hyperbolic secant), csch (hyperbolic cosecant), and coth (hyperbolic cotangent). Hyperbolic functions have applications in physics, especially in areas involving waves and electromagnetism (Saxena, 2011).
6. **Rational Functions:** These are the ratio of two polynomial functions. They are of the form f(x) = p(x)/q(x), where p(x) and q(x) are polynomials. Rational functions have a numerator and a denominator and may have vertical asymptotes where the function approaches infinity or negative infinity.

5.2.2. Function Operations and Compositions

1. **Addition, Subtraction, Multiplication, and Division of Functions:** Functions can be combined using arithmetic operations. Given two functions f(x) and g(x), their sum (f(x) + g(x)), the difference (f(x) – g(x)), product (f(x) × g(x)), and quotient (f(x)/g(x), where g(x) is not equal to zero) can be calculated. These operations are valid for combining and

manipulating functions to solve equations and model complex relationships.

2. **Composition of Functions:** Function composition is the process of applying one function to the result of another function. If we have two functions, f(x) and g(x), the composition f(g(x)) represents applying g(x) to the input x and then using the output of g(x) as the input to f(x). The order of composition matters and not all functions can be composed (Saxena, 2011).
3. **Inverse Functions and Their Properties:** The inverse function of a given function f(x), denoted as f^(–1)(x), undoes the operation of f(x). For a function to have an inverse, it must be one-to-one (injective), meaning each element in the domain has a unique component of the codomain. The inverse function undoes the role of the original function, mapping the output values back to the original input values. The inverse of a function f(x) is represented as y = f^(–1)(x).

5.2.3. Transformations of Functions

1. **Horizontal and Vertical Translations:** These involve shifting a function horizontally or vertically along the coordinate axes. A horizontal translation of h units to the right is represented by f(x – h), while a translation of k units upward is represented by f(x) + k. These translations change the position of the graph without altering its shape.
2. **Reflections and Stretches:** Reflections involve flipping the graph of a function across the x-axis or y-axis. A review across the x-axis is represented by –f(x), while a reflection across the y-axis is denoted by f(–x). Stretches involve multiplying the function by a constant 'a' to stretch or compress the graph vertically. For example, AF (x) represents a vertical stretch if a > 1 or compression if 0 < a < 1.
3. **Combining Transformations:** Multiple transformations can be connected to modify the graph of a function. For example, a function can undergo a horizontal translation, a reflection across the y-axis, and a vertical stretch. Understanding how these transformations affect the graph is crucial for analyzing and interpreting functions in various contexts (Luke, 2014).

5.3. FUNCTION PROPERTIES AND GRAPHS

5.3.1. Continuity and Differentiability

1. **Definition of Continuity:** A function is continuous at a point c if three conditions are met:
- The process is defined at c.
- The limit of the function as x approaches c exists.
- The limitation and the function value at c are equal, i.e., lim[x→c] f(x) = f(c).
2. **Continuity of Elementary Functions:** Elementary functions such as polynomials, trigonometric functions, exponential functions, and logarithmic functions are continuous over their entire domains. This property makes them valuable in modeling real-world phenomena and performing mathematical operations.
3. **Differentiability and the Derivative of a Function:** A function f(x) is considered differentiable at point c if the derivative f'(c) exists. Geometrically, the derivative represents the tangent line's slope to the function's graph at point c. If a function is differentiable at every point in its domain, it is considered differentiable over that domain (Saxena, 2011).
4. **Differentiation Rules for Elementary Functions:** Differentiation rules enable us to find the derivative of more complex functions by combining the results of elementary operations. For example, the power, product, quotient, and chain rules are essential tools for finding polynomial, trigonometric, exponential, and composite function derivatives.

5.3.2. Symmetry and Asymptotes

1. **Symmetry Properties of Functions:** Functions may exhibit various types of symmetry. A function f(x) is even if f(–x) = f(x), meaning it is symmetric concerning the y-axis. A function is odd if f(–x) = –f(x), indicating symmetry about the origin. Some functions, such as polynomials with odd degrees, exhibit both even and odd symmetry (Luke, 2014).
2. **Horizontal and Vertical Asymptotes:** Asymptotes are lines that a function's graph approaches but never intersects. A horizontal

asymptote is a horizontal line that the role approaches as x approaches positive or negative infinity. Vertical asymptotes occur when the function approaches infinity or negative infinity as x approaches a specific value.

3. **Oblique Asymptotes:** Also known as slant asymptotes, occur when a function approaches a straight line as x approaches positive or negative infinity. Unlike horizontal or vertical asymptotes, oblique asymptotes are represented by linear $y = mx + b$ equations.

5.3.3. Extrema and Inflection Points

1. **Local and Global Extrema:** Extrema refers to a function's maximum and minimum points. A local maximum is a point where the process is more significant than its nearby values, and a local minimum is a point where the function is smaller than its nearby values. Global extrema are the function's absolute maximum and minimum values over its entire domain.
2. **Finding Critical Points:** Critical points are the values of x where the function's derivative is either zero or undefined. The function may have local extrema or points of inflection at critical points (Saxena, 2011).
3. **Inflection Points and Concavity:** Inflection points are points on the graph of a function where the concavity changes. At an inflection point, the function transitions from concave up to concave down or vice versa. The sign of the second derivative of the function determines concavity.

5.3.4. Graphs of Special Functions

1. **Graphs of Trigonometric Functions:** Trigonometric functions, such as sine, cosine, tangent, cosecant, secant, and cotangent, have distinct periodic properties. Their charts oscillate between specific values, and they have unique symmetries. Trigonometric functions are fundamental in modeling rare phenomena like waves and oscillations.
2. **Graphs of Exponential and Logarithmic Functions:** Exponential functions have exponential growth or decay, leading to rapid increases or decreases. Their charts exhibit asymptotic

behavior as they approach the x-axis. Logarithmic functions, the inverse of exponential functions, display opposite behavior, with slow growth or decay as the input values increase (Luke, 2014).

3. **Graphs of Rational Functions:** Rational functions are characterized by dividing two polynomial functions. Their charts may have vertical asymptotes where the function approaches infinity or negative infinity. Rational functions can also have horizontal asymptotes as the numerator and denominator polynomials' degree comes to the same value.
4. **Graphs of Hyperbolic Functions:** Hyperbolic functions have similar forms to trigonometric functions but are defined using the hyperbola. Their unique symmetries and exponential properties make them valuable in physics, especially in modeling waves and electrical circuits (Saxena, 2011).

5.4. FUNCTION ANALYSIS AND APPLICATIONS

5.4.1. Limits and Continuity

1. **Limit Definition and Properties:** The limit of a function f(x) as x approaches a specific value c represents the function's behavior as x gets arbitrarily close to c. The limit is denoted as lim[x→c] f(x). It can exist even if the function is not defined at c. Properties of limitations, such as the sum, product, and quotient rules, allow for evaluating complex limits based on the limits of more straightforward functions.
2. **Evaluating Limits of Functions:** Evaluating limits involves finding the value a function approaches as the input value gets arbitrarily close to a given point. This process helps analyze the behavior of functions at specific points and understand their continuity and behavior around critical issues.
3. **Continuity and Limits:** The concept of limits is closely related to the continuity of a function. A function is continuous at a point c if the limit of the function as x approaches c exists and is equal to the function value at c. Continuity ensures no gaps, jumps, or breaks in the function's graph. Functions are continuous or discontinuous based on their behavior at specific points or intervals.

5.4.2. Applications of Functions in Physics and Engineering

1. **Modeling Physical Phenomena with Functions:** Functions play a central role in modeling various physical phenomena. In physics, functions are used to describe the motion of objects, the behavior of waves, the distribution of particles, and the relationship between physical quantities.
2. **Functions in Classical Mechanics:** Classical mechanics relies heavily on mathematical functions to describe the motion of objects under the influence of forces. Functions are used to model the position, velocity, acceleration, and energy of particles and bodies (Luke, 2014).
3. **Electrical Engineering Applications of Functions:** In electrical engineering, functions are used to model electrical circuits and systems. They are crucial for understanding the behavior of electrical components, analyzing signal processing systems, and designing control systems.
4. **Control Systems and Functions:** Functions are employed in control systems engineering to represent the input-output relationships of dynamic systems. Control theory uses functions to design controllers that regulate the behavior of systems and achieve desired responses (Saxena, 2011).

5.4.3. Functions in Economics and Finance

1. **Economic Modeling Using Functions:** Functions are fundamental in financial modeling, representing the relationships between economic variables. They are used to model production, consumption, investment, and other financial processes.
2. **Supply and Demand Functions:** Supply and demand functions are essential tools in microeconomics. They illustrate the number of goods or services producers are willing to supply and the amount consumers are eager to demand at different price levels.
3. **Utility Functions and Consumer Behavior:** Utility functions model consumer preferences and decision-making in microeconomics. Utility functions represent how consumers derive satisfaction from goods and services, helping economists analyze consumer behavior.

4. **Financial Applications of Functions:** Functions are widely used in finance for various purposes. They are used to model financial markets, analyze investment growth, calculate interest rates, and determine risk in financial instruments.

5.4.4. Functions in Computer Science and Data Analysis

1. **Functions in Algorithms and Programming:** In computer science, functions are a fundamental concept in programming. Functions allow programmers to encapsulate code into reusable blocks, improving code modularity and readability (Luke, 2014).
2. **Data Transformation Using Functions:** Functions are critical in data transformation and manipulation. Data can be transformed using mathematical functions to scale, normalize, or preprocess data for analysis or visualization.
3. **Machine Learning and Functions:** Machine learning algorithms heavily rely on functions to model relationships between input data and output predictions. Functions define the structure and behavior of machine learning models, enabling them to learn from data and make accurate predictions.
4. **Functions in Data Visualization:** Data visualization often involves mathematical functions to create graphical data representations. Functions are used to create charts, graphs, and plots to help interpret and communicate complex data patterns effectively (Saxena, 2011).

5.4.5. Solving Equations Involving Functions

1. **Algebraic Methods for Solving Equations:** Functions are used in solving algebraic equations. Techniques such as factoring, completing the square, and using the quadratic formula are applied to find the roots or solutions of polynomial equations.
2. **Numerical Methods for Solving Equations:** In cases where equations cannot be solved analytically, numerical methods approximate solutions. Techniques like the bisection, Newton, and secant methods are used to find numerical solutions to equations.
3. **Applications in Engineering and Science:** Solving equations involving functions is crucial in engineering and scientific fields.

Engineers use equation solving to design structures, analyze systems, and optimize performance. Scientists use equation solving to model physical phenomena, interpret experimental data, and derive conclusions from mathematical models.

5.5. SPECIAL FUNCTIONS AND ADVANCED TOPICS

5.5.1. Special Elementary Functions

1. **Gamma Function (Γ(x)):** It is a continuous extension of the factorial function to real and complex numbers. It is defined as Γ(x) = ∫[0 to ∞] t^(x–1) × e^(–t) dt. For positive integer values of x, the gamma function simplifies to Γ(x) = (x–1)! However, the gamma function is also defined for non-integer values, making it a powerful tool for interpolation and approximation. The gamma function has applications in various areas, such as probability, combinatorics, and number theory. For example, the gamma function is used to compute combinatorial coefficients, such as binomial coefficients and Euler numbers. Additionally, it appears in the solution of various integral equations, such as evaluating definite integrals (Luke, 2014).
2. **Beta Function (B(x, y)):** It is closely related to the gamma function and is defined as B(x, y) = Γ(x) × Γ(y)/Γ(x + y). It plays a significant role in statistics and probability theory, particularly in defining the beta distribution. The beta distribution is widely used to model random variables that are bounded between 0 and 1, such as proportions and probabilities. For example, in Bayesian statistics, the beta distribution is commonly used as a conjugate prior to the binomial distribution, allowing for efficient parameter estimation and updating of probabilities.
3. **Error Function (erf(x)):** It is defined as erf(x) = (2/√π) ∫[0 to x] e^(–t^2) dt. The error function arises in probability theory, statistics, and physics, particularly in describing the cumulative distribution function (CDF) of a standard normal distribution. It calculates the probabilities of events occurring within a certain range in a normal distribution. For example, the error function is utilized to find the probability that a normally distributed random variable falls within a specified number of standard deviations from the mean. The error function is also crucial in solving

differential equations, particularly in heat conduction problems and diffusion processes, where it represents a system's spatial and temporal evolution (Saxena, 2011).

4. **Dirac Delta Function (δ(x)):** It is a distribution rather than a conventional function. It is defined as zero for all values of x except at x = 0, where it becomes infinite while satisfying the condition that the integral of the delta function over the entire natural line is equal to one. The Dirac delta function is used in physics and engineering to model point sources, impulse responses, and singularities in differential equations. For example, in electrical engineering, the Dirac delta function describes a unit impulse in a circuit, representing an instantaneous, infinitely large current or voltage at a specific time.

5.5.2. Orthogonal Functions

1. **Orthogonal Functions and Inner Products:** Orthogonal functions are mutually perpendicular when viewed as vectors in a function space. The inner product of two orthogonal functions is zero, except when the functions are the same, where the inner product is nonzero. Orthogonal functions have essential properties that allow them to form a basis for function spaces, making them powerful tools for solving differential equations and representing complex functions (Luke, 2014).
2. **Legendre Polynomials (Pn(x)):** These are a family of orthogonal polynomials that have applications in physics, particularly in solving partial differential equations, quantum mechanics, and representing potential fields. They are solutions to Legendre's differential equation and are defined by the Rodrigues formula: Pn(x) = (1/2^n) × (d^n/dx^n) [(x^2 – 1)^n]. Legendre polynomials have unique properties, such as symmetry and recurrence relations, making them valuable in solving problems involving spherical harmonics and boundary value problems in electrostatics and fluid dynamics.
3. **Hermite Polynomials (Hn(x)):** These are another set of orthogonal polynomials with applications in quantum mechanics, particularly in solving the Schrödinger equation for the harmonic oscillator potential. They are solutions to Hermite's differential equation and are defined by the Rodrigues formula: Hn(x) = (–1)^n × e^(x^2) × (d^n/dx^n) [e^(–x^2)]. Hermite polynomials

describe the energy levels of the quantum harmonic oscillator and appear in statistical mechanics, particularly in modeling Brownian motion and quantum statistical distributions (Saxena, 2011).

4. **Bessel Functions (Jn(x)):** These are solutions to Bessel's differential equation and are essential in solving wave phenomena, such as heat conduction, sound waves, and electromagnetic waves. They appear in various physical phenomena involving waves and oscillations, such as vibrating strings, circular membranes, and diffraction patterns. Bessel functions have unique properties, including differentiability and orthogonality, making them indispensable in solving boundary value problems in cylindrical and spherical geometries.

5.5.3. Applications of Special Functions

1. **Wave Phenomena and Bessel Functions:** Bessel functions are fundamental in studying wave phenomena. In the case of heat conduction, the Bessel functions appear as solutions to the heat equation in cylindrical coordinates. They describe the temperature distribution in a cylindrical object with a constant heat source at the center. Bessel functions describe the pressure distribution in a cylindrical resonator for sound waves. Additionally, in diffraction patterns, Bessel functions describe the intensity distribution of waves passing through a circular aperture (Luke, 2014).
2. **Quantum Mechanics and Hermite Polynomials:** Hermite polynomials play a vital role in quantum mechanics, particularly in solving the Schrödinger equation for the harmonic oscillator potential. The energy levels of the quantum harmonic oscillator are quantized and proportional to the Hermite polynomials, making them central to understanding the quantization of energy in quantum systems. In the context of quantum states, Hermite polynomials appear in the description of wave functions and the probability densities of quantum particles confined in harmonic potentials.
3. **Statistical Mechanics and Partition Functions:** Special functions, such as the gamma and beta functions, appear in statistical mechanics, specifically in calculating partition functions. Partition functions are crucial in determining the thermodynamic properties of systems and predicting macroscopic

behavior from microscopic properties. For example, the partition function of a quantum system is related to the system's free energy and can be used to calculate the system's average energy and entropy (Saxena, 2011).

5.6. COMPLEX FUNCTIONS AND ANALYTICITY

5.6.1. Complex Numbers and Functions

1. **Complex Number Representation:** Complex numbers are represented as z = a + bi, where a and b are real numbers, and i is the imaginary unit (i^2 = –1). The fundamental part a is denoted as Re(z), and the imaginary part b as Im(z). The complex plane is used to visualize complex numbers, with the real axis representing the actual numbers and the imaginary axis representing the imaginary numbers.
2. **Complex Arithmetic and Operations:** Complex arithmetic involves addition, subtraction, multiplication, and division of complex numbers. These operations are performed separately for the real and imaginary parts. For example, the sum of two complex numbers z1 = a1 + b1i and z2 = a2 + b2i is z1 + z2 = (a1 + a2) + (b1 + b2)i.
3. **Complex Conjugate and Modulus:** The complex conjugate of a complex number z = a + bi is given by z– = a – bi*. The modulus (or absolute value) of z is defined as |z| = √(a^2 + b^2). The modulus of a complex number represents its distance from the origin in the complex plane.

5.6.2. Analytic Functions

1. **Definition of Analyticity:** A complex function f(z) is said to be analytic (or holomorphic) at a point z0 if it is differentiable in some neighborhood of z0. This implies that the derivative f'(z0) exists; therefore, f(z) has a well-defined tangent at z0. A function that is analytic throughout a region is called an entire function (Luke, 2014).
2. **Cauchy-Riemann Equations:** Analyticity is closely related to the Cauchy-Riemann equations, which provide necessary and sufficient conditions for a function to be analytic. For a function

f(z) = u(x, y) + iv(x, y) to be logical, its real and imaginary parts, u(x, y) and v(x, y), must satisfy the Cauchy-Riemann equations: $\partial u/\partial x = \partial v/\partial y$ and $\partial u/\partial y = -\partial v/\partial x$.

3. **Harmonic Functions:** If a complex function f(z) = u(x, y) + iv(x, y) is analytic, its real and imaginary parts, u(x, y) and v(x, y), are harmonic functions. A function is harmonic if it satisfies Laplace's equation: ∇^2 u = 0, where ∇^2 is the Laplacian operator. Harmonic functions arise in various physical and engineering problems, including heat conduction and fluid flow (Saxena, 2011).

5.6.3. Complex Exponential and Logarithmic Functions

1. **Euler's Formula and its Implications:** Euler's formula is a fundamental result in complex analysis, given by e^(iθ) = cos(θ) + isin(θ)*, where θ is the argument (or angle) of the complex number. This formula establishes a profound connection between exponential, trigonometric, and imaginary functions. Euler's formula is essential in expressing complex numbers in polar form and simplifying complex arithmetic.
2. **Complex Logarithm and Multivaluedness:** The complex logarithm, denoted as Log(z), is the inverse of the complex exponential function. However, the complex logarithm is multivalued, meaning that for a given complex number z, there are infinitely many possible logarithmic values. This is due to the periodicity of the exponential function with a period of $2\pi i$. A branch cut is introduced to define a single-valued complex logarithm, usually along the negative real axis.
3. **Branch Cuts and Principal Value:** A branch cut is a curve in the complex plane along which the complex function becomes discontinuous. The branch cut is introduced for the complex logarithm to ensure the function is single-valued. The principal value of the complex logarithm, denoted as Log(z), is defined by taking the logarithm value with the argument in the range $(-\pi, \pi]$ (Saxena, 2011).

5.6.4. Complex Trigonometric and Hyperbolic Functions

Complex trigonometric and hyperbolic functions are extensions of their natural counterparts to complex numbers. They play a significant role in complex analysis and have various applications in physics, engineering, and other scientific disciplines (Luke, 2014).

1. **Complex Sine and Cosine Functions:** The exponential function and Euler's formula define the complex sine and cosine functions. For a complex number z = x + iy, where x and y are real numbers, the complex sine and cosine functions are given by:

Complex Sine (sin(z)) = (e^(iz) – e^(–iz))/(2i)

Complex Cosine (cos(z)) = (e^(iz) + e^(–iz))/2

These complex trigonometric functions have periodicity in the complex plane, just like their real counterparts. For any complex number z, sin(z + 2π) = sin(z) and cos(z + 2π) = cos(z). The complex sine and cosine functions are functions used to solve problems involving oscillatory behavior and wave phenomena in complex analysis.

2. **Complex Hyperbolic Functions:** These are analogs of the complex sine and cosine functions, obtained by replacing the trigonometric functions with hyperbolic functions in their definitions. For a complex number z = x + iy, the complex hyperbolic functions are given by:

Complex Hyperbolic Sine (sinh(z)) = (e^(z) – e^(–z))/2

Complex Hyperbolic Cosine (cosh(z)) = (e^(z) + e^(–z))/2

The complex hyperbolic functions also exhibit periodicity in the complex plane and are entire functions. The complex hyperbolic functions are used to model problems involving exponential growth and decay in complex analysis.

3. **Relationship to Real Trigonometric and Hyperbolic Functions:** The complex trigonometric and hyperbolic functions are closely related to their real counterparts. When the imaginary part of a complex number z is zero, the complex trigonometric and hyperbolic functions reduce to the real trigonometric and hyperbolic functions, respectively. For example, if z = x (where x is real), then sin(x) = sin(z), cos(x) = cos(z), sinh(x) = sinh(z), and cosh(x) = cosh(z).

Furthermore, Euler's formula allows us to express complex trigonometric functions in terms of fundamental trigonometric functions and vice versa. For instance, we can write sin(z) and cos(z) in terms of sinh(z) and cosh(z) using Euler's formula (Luke, 2014).

5.7. APPLICATIONS OF COMPLEX FUNCTIONS

Complex functions have various applications in physics, engineering, and scientific disciplines. This section explores the practical use of complex analysis in solving real-world problems and understanding complex phenomena.

5.7.1. Complex Analysis in Physics and Engineering

1. **Application of Complex Functions in Electrical Engineering:** Complex functions are extensively used in electrical engineering to analyze and design circuits and systems. They provide a powerful tool for understanding the behavior of circuits with alternating currents (AC) and impedance. For example, a capacitor or inductor impedance in an AC circuit is a complex quantity, and complex functions help calculate voltage and current phase differences, power factors, and resonance frequencies in AC circuits.
2. **Resonance Phenomena and Complex Frequency Analysis:** Resonance is a phenomenon where a system exhibits maximum response to an external force at a particular frequency. Complex functions analyze resonance in various methods, including mechanical oscillators, electrical circuits, and electromagnetic structures. The concept of complex frequency analysis allows engineers and physicists to understand resonant systems' stability and dynamic behavior (Saxena, 2011).
3. **Fluid Dynamics and Potential Flow:** Complex functions play a crucial role in fluid dynamics, particularly in studying potential flow. In possible flow, fluid motion is irrotational and can be described by a complex possible function. The use of complex analysis in fluid dynamics enables the study of flow patterns around objects, such as airfoils, and the prediction of lift and drag forces on them.

5.7.2. Complex Integration and Residue Theorem

1. **Contour Integration in the Complex Plane:** Complex integration involves the integration of complex functions along curves in the complex plane. Contour integration is a powerful technique used to evaluate complex integrals and compute residues of functions.

2. **Cauchy's Integral Theorem and Formula:** Cauchy's integral theorem states that if a function is analytic within a connected region and along its boundary, then the integral of the function along any closed curve within that region is zero. Cauchy's integral formula establishes a connection between complex integration and the values of an analytic process within a closed contour.
3. **Residue Theorem and its Applications:** The residue theorem is a significant result in complex analysis, stating that the integral of an analytic function around a closed contour equals $2\pi i$ times the sum of the residues of the process at its isolated singularities within the shape. The residue theorem is invaluable in evaluating complex integrals, particularly contour integration and computing real integrals using complex methods (Luke, 2014).

5.7.3. Mapping with Complex Functions

1. **Conformal Mapping and Its Properties:** Conformal mapping is a particular type of mapping using complex functions that preserves angles locally. Conformal maps transform one complex domain into another while maintaining the shape and angles of figures. Conformal mapping has significant applications in engineering and physics, particularly in the design of aerodynamic shapes, heat transfer analysis, and fluid flow modeling.
2. **Applications of Conformal Mapping in Engineering and Physics:** Conformal mapping has many practical applications. For example, in electrical engineering, it is used to map complex geometries to simpler shapes for analysis, such as solving Laplace's equation for electrostatic problems. In fluid dynamics, it is used to study flow around complex objects by mapping the flow domain to a simpler one. Conformal mapping also plays a role in designing electronic circuits, antennas, and even cartography for map projections (Saxena, 2011).
3. **Transforming Shapes Using Complex Functions:** Complex functions can transform shapes and regions in the complex plane. For instance, the mapping function w = z^2 squares the complex number z, changing a circle in the z-plane to a cardioid shape in the w-plane. Transformations like this help visualize complex functions and understand how they affect bodies and regions (Luke, 2014).

5.8. NUMERICAL METHODS FOR FUNCTIONS

Numerical methods play a vital role in approximating and solving difficult or impossible functions to evaluate analytically. This section explores various numerical techniques for comparing functions, performing numerical integration, and solving differential equations.

5.8.1. Approximation of Functions

1. **Taylor Series Expansion:** It is a method for approximating a function using its derivatives evaluated at a specific point. The Taylor series representation of a function f(x) centered at x = a is given by:

$$f(x) = f(a) + f'(a)(x - a) + f''(a)(x - a)^2/2! + f'''(a)(x - a)^3/3! + \ldots$$

We can obtain a polynomial approximation of the function around the point x = a by truncating the series to a finite number of terms. The Taylor series is particularly useful for approximating smooth functions and evaluating functions at points where their analytic expressions are difficult to compute.

2. **Polynomial Interpolation:** It is a numerical method for approximating a function using a polynomial that passes through a set of given data points. Given n+1 data points (x0, y0), (x1, y1),…, (xn, yn), the polynomial of degree n that interpolates these points is unique and can be determined using various interpolation techniques, such as Lagrange interpolation or Newton's divided differences. Polynomial interpolation is widely used in numerical analysis and data fitting applications.
3. **Least Squares Fitting:** It is a method used to approximate a function by finding the best-fitting curve or surface that minimizes the sum of the courts of the differences between the actual data points and the function's values. This approach is often used when the data is noisy or contains errors. The method can be applied to linear and nonlinear functions and is widely used in regression analysis and curve fitting (Saxena, 2011).

5.8.2. Numerical Integration of Functions

1. **Newton-Cotes Formulas (Trapezoidal and Simpson's Rule):** These are numerical techniques for approximating the definite integral of a function. The Trapezoidal rule divides the interval of

integration into small trapezoids and approximates the area under the curve as the sum of the areas of these trapezoids. Simpson's rule further improves the accuracy by comparing the function as a quadratic polynomial within each interval and using the site under each parabolic segment to approximate the integral (Luke, 2014).

2. **Gaussian Quadrature:** It is a numerical method for numerical integration that provides more accurate results than Newton-Cote formulas for various function evaluations. It uses a weighted sum of function values at specific points (called quadrature points) within the integration interval. Gaussian quadrature is designed to integrate polynomials of particular degrees strictly and is widely used in engineering and scientific applications.
3. **Monte Carlo Integration:** It is a probabilistic numerical method for approximating integrals using random sampling. It involves generating random points within the integration region and using the average function values at these points to reach the integral. Monte Carlo integration benefits high-dimensional integrals and problems where other numerical integration methods may be computationally expensive.

5.8.3. Solving Differential Equations with Functions

1. **Numerical Methods for Ordinary Differential Equations (ODEs):** ODEs involve derivatives of a function concerning a single independent variable. Numerical methods for solving ODEs include finite difference, Runge-Kutta, and multistep processes like Adams-Bashforth and Adams-Moulton. These methods approximate the solution of the ODE by discretizing the domain and calculating the function values at each time step (Saxena, 2011).
2. **Finite Difference Methods:** These are numerical techniques for solving differential equations by approximating derivatives using difference quotients. The results are expressed as the difference between function values at neighboring points by dividing the domain into small intervals. This approach is beneficial for solving partial differential equations and other boundary value problems.

3. **Euler's Method and Higher-Order Methods:** Euler's method is a basic numerical technique for solving first-order ODEs. It approximates the solution by iteratively evaluating the derivative at each step and updating the function value accordingly. Higher-order methods, such as the Runge-Kutta methods, use multiple function evaluations and weighted combinations of slopes to achieve higher accuracy and stability (Luke, 2014).

5.9. ADVANCED TOPICS IN FUNCTIONAL ANALYSIS

Functional analysis is a branch of mathematics that deals with vector spaces of functions and the analysis of function spaces. This section explores advanced topics in available analysis, including Banach and Hilbert spaces, operator theory, and their applications in quantum mechanics.

5.9.1. Banach and Hilbert Spaces

1. **Definition and Properties of Banach Spaces:** A Banach space is an utterly normed vector space, meaning that every Cauchy sequence converges to a limit within the area. Banach rooms have a norm, which measures the size or length of vectors in the space. Critical properties of Banach spaces include the triangle inequality and the completeness property. Banach areas include the opening of continuous functions and integrable functions.
2. **Hilbert Spaces and Inner Products Spaces:** Hilbert spaces are a particular type of Banach space with an inner product defined on them. The inner product is a generalization of the dot product and allows us to measure the angle between vectors and define notions of length and orthogonality. Hilbert spaces have the additional property of being complete concerning the norm induced by the inner product. Examples of Hilbert spaces include the space of square-integrable functions and Euclidean spaces (Saxena, 2011).
3. **Orthogonal Bases and Fourier Series:** In Hilbert spaces, orthogonal bases play a crucial role in representing functions as infinite series. The Fourier series is an essential application of functional analysis, where functions can be defined as a sum of orthogonal trigonometric functions. The completeness and orthogonality properties of the basis allow for efficient

approximations of functions and solving differential equations through the separation of variables.

5.9.2. Operator Theory and Linear Transformations

1. **Linear Operators and Their Properties:** In functional analysis, linear operators are mappings between function spaces that preserve the vector space structure and satisfy linearity properties. They are widely used to model various physical and mathematical processes, such as differentiation, integration, and convolution. Linear operators have essential properties, including linearity, continuity, and boundedness.
2. **Eigenvalues and Eigenvectors:** In the context of operator theory, eigenvalues, and eigenvectors are crucial concepts. For a linear operator T, an eigenvalue is a scalar λ for which a nonzero vector v (the eigenvector) exists such that T(v) = λv. The study of eigenvalues and eigenvectors is central to understanding the behavior and stability of linear systems (Luke, 2014).
3. **Spectral Theorem and Applications:** The spectral theorem is a fundamental result in functional analysis that provides a powerful representation of self-adjoint operators (operators equal to their adjoint). It states that every self-adjoint operator on a Hilbert space has a spectral decomposition, expressed as a sum of projections onto eigenspaces associated with distinct eigenvalues. The spectral theorem has significant applications in quantum mechanics and signal processing.

5.9.3. Functional Analysis in Quantum Mechanics

1. **Function Spaces in Quantum Mechanics:** In quantum mechanics, a wave function, which belongs to a function space, describes the state of a quantum system. The most commonly used function spaces in quantum mechanics are Hilbert spaces, such as the space of square-integrable functions. The properties of function spaces are crucial in characterizing quantum states and the evolution of quantum systems (Saxena, 2011).
2. **Wavefunctions and Probability Density:** The wavefunction of a quantum system encodes the probability amplitude of finding the system in a particular state. The square modulus of the

wavefunction represents the probability density, and the integral of the probability density over the entire space gives the total probability of finding the system in any state.

3. **Hermitian Operators and Observables:** Hermitian operators in the Hilbert space represent observations in quantum mechanics. Hermitian operators have real eigenvalues, and their eigenvectors correspond to the states where the observables are well-defined. The observables' Measurement outcomes are the corresponding operator's eigenvalues, and the probability of obtaining a specific measurement outcome is determined by the square modulus of the projection of the wavefunction onto the corresponding eigenvector (Luke, 2014).

In conclusion, mathematical functions are fundamental concepts in mathematics with broad applications in science, engineering, and various other fields. They serve as crucial tools for understanding and modeling relationships between variables, allowing us to analyze and predict complex phenomena in the natural world (Saxena, 2011).

Throughout this chapter, we explored the different types of mathematical functions, ranging from elementary operations like polynomials and trigonometric functions to special ones like the gamma and Bessel functions. We also delved into the properties and graphs of functions, understanding concepts like continuity, differentiability, and extrema.

Moreover, we discussed advanced functional analysis topics, including Banach and Hilbert spaces, operator theory, and their applications in quantum mechanics. These concepts provide a deeper understanding of function spaces and linear operators, enabling us to study complex systems and solve differential equations.

The significance of functions in various fields cannot be overstated. In physics, functions help model physical phenomena; in engineering, they aid in designing systems and analyzing complex structures. In economics and finance, functions are essential for economic modeling and financial analysis, and in computer science, they are utilized for data transformation and machine learning.

In the future, further research and exploration of mathematical functions can lead to discoveries and advancements in various disciplines. The possibilities are vast, from studying special functions in number theory to exploring novel applications in emerging fields like quantum computing and data science (Luke, 2014).

CHAPTER 6

Linear Algebra in Mathematics

Contents

Linear algebra is fundamental to the study of mathematics. Throughout linear algebra, the notions of vectors and linear functions take center stage. It is a basic notion that serves as the foundation for the great bulk of mathematics. As a core idea, linear algebra is heavily emphasized in the current geometry curriculum. Planes, lines, and object rotations are among of the most basic things that it may explain in engineering and the physical sciences. Furthermore, the high degree of processing efficiency allows us to correctly show a wide range of natural phenomena. In this chapter, we'll go over the essentials of linear algebra, such as its structure, common problems, fundamental linear equations, and useful applications. We will also go over some basic linear equations. Linear combinations are the fundamental focus of linear algebra. Before applying any linear transformations, it is required to do extensive study on vector spaces, lines, planes, and particular mappings. Its basic pieces are linear functions, matrices, and vectors. An investigation of linear equation sets and their ability to undergo transformations (Poole, 2014).

In Algebra, a linear equation is used. In its most frequent form, the linear equation looks like this:

(ax1)(ax2) + b = a(nx)n

Here, the coefficients of the equation are denoted by the letter a. The Xs represent the situation's unknowns. Constant values are indicated by the letter B. A "system" is a set of linear algebraic equations. Using matrices, it is feasible to obtain a solution to the system of equations. Linear functions such as a(a1x1)+(a2x2)+(a3x3)+(a4x) = a(a1x1)+(a2x2)+(a3x3)+(a4x) and others are supported. The chapter "Rectangular Arrays" in The Nine Chapters on the Mathematical Art, a historic Chinese mathematical treatise, is named "Rectangular Arrays." This chapter contains the modern approach of Gaussian elimination for concurrently solving linear equations (by the use of counting rods). There are two to five equations in each of the eighteen separate problems. René Descartes' application of coordinates to the study of geometry in 1637 is regarded as the beginning point for the creation of linear equation systems in Europe. Lines and planes are truly represented by linear equations in this new geometry, which is now known as Cartesian geometry, and finding the place where two lines cross involves the solution of a system of linear equations (Strang, 2022). Leibniz' important work on determinants, finished in 1693, paved the stage for the first systematic procedures for solving linear equations to be devised. They were utilized by Gabriel Cramer in 1750 to develop what is now known as Cramer's method for finding explicit solutions to linear equations. To find explicit solutions to linear equations, Cramer's rule is utilized. Gauss' elimination approach was first hailed as a substantial accomplishment in the science of geodesy. "Theory of Extension," published in 1844 by Hermann Grassmann, was the first location where essential original notions that would later become known as linear algebra were presented. In the year 1848, James Joseph Sylvester coined the Latin term "matrix," which may be rendered as "womb."

The study of the complex plane and the clarification of the notions associated with it were significant factors to the creation of linear algebra. Despite the fact that the difference between the integers w and z in mathbb C is indicated by w–z, the length and orientation of the line segments wz and 0(w z) are the same. Each component works in tandem with the others. In 1843, W. R. Hamilton created the mathbb H quaternions system, which functions in four dimensions. A "vector" is a point in space defined by the equation v = xi + yj + zk. This is an explanation of the word "vector." Furthermore, the segment pq is created by subtracting the quaternions p and q from each other. Similarly, more sophisticated number systems often use the idea of a linear space without a basis (Poole, 2014). Arthur Cayley discovered matrix multiplication and the inverse matrix in the year 1856. These breakthroughs cleared the path for the existence of the generic linear group. The application of the group representation technique has enabled the clarification of complex and hypercomplex numbers. Cayley, for example, utilized a single letter to describe the concept of a matrix that she was employing as a grouping technique. In answer to an inquiry on the relationship between matrices and determinants, he said, "There would be much to say about this theory of matrices, which seems to come before the theory of determinants." Although Benjamin Peirce was the first to create Linear Associative Algebra in 1872, it was his son, Charles Sanders Peirce, who subsequently improved on his father's work. The 1873 publication A Treatise on Electricity and Magnetism is recognized as being the first to propose the field theory of forces and its representation in terms of differential geometry. On the telegraph, further clarity was required. Linear algebra is a differential geometry discipline that treats manifolds as tangent spaces and performs operations on them on a two-dimensional plane. Lorentz transformations were crucial in the creation of linear algebra because they represent electromagnetic symmetries of spacetime. Lorentz found them, which explains why they were so successful (Poole, 2014). A theory of linear transformations of vector spaces with a restricted number of dimensions has already been created by the year 1900. In the year 1888, Peano presented the first formal and detailed definition of a vector space. Many earlier notions and procedures were abstracted into algebra in the early part of the 20th century, resulting in the formation of the modern topic of linear algebra as a field of study. As computing power has increased, linear algebra has grown in importance as a modeling and simulation tool. This has inspired research into more efficient Gaussian elimination and matrix decomposition algorithms (Strang, 2022).

6.1. LINEAR ALGEBRA TOPICS

There are a number of topics studied under linear algebra. These topics are very broad and extensive. Some of the topics are studies below:

6.1.1. Euclidean Vector Spaces

The primary space of geometry, Euclidean space, is supposed to be a representation of the world as it really exists. It began with the three-dimensional space of Euclidean geometry, as described in Euclid's Elements. However, according to recent mathematical advancements, Euclidean spaces may have any positive integer number n, and these spaces are known as Euclidean n-spaces. When n is 1 or 2, the lines and planes are said to be Euclidean. In modern mathematics and science, the word "Euclidean" is used to distinguish Euclidean spaces from other kinds of spaces (Poole, 2014) (Figure 6.1).

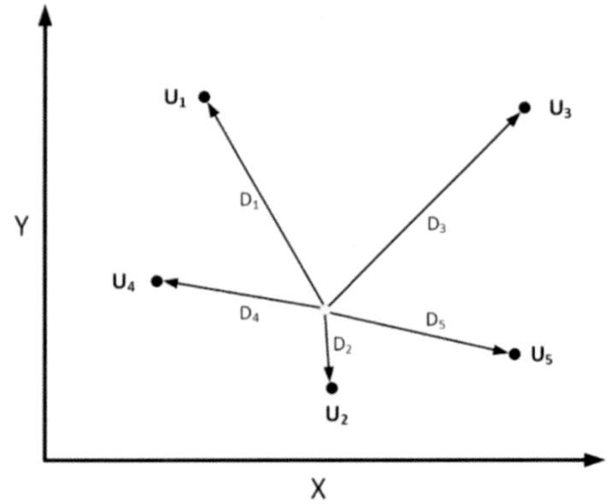

Figure 6.1. Two-dimensional vector space representation of the Euclidean classifier where the distance function between each class and a sub-area is analyzed.

Source: Cortés-Polo, Jimenez, Calle-Cancho, & González-Sánchez (2019); Scientific Figure on ResearchGate.

The notion of Euclidean space was developed by ancient Greek geometers in attempt to create a model that correctly mirrored the real world. In his book Elements, Euclid, an ancient Greek mathematician, gathered the work of his contemporaries and achieved the major breakthrough of demonstrating all spatial qualities as theorems. He accomplished this by starting with a few basic features, or postulates, which were either self-evident (such as the notion that there is precisely one straight line traveling between two locations) or looked to be difficult to establish (the parallel postulate).

As a response to the development of non-Euclidean geometries, the prior postulates were re-formalized to define Euclidean spaces using axiomatic theory towards the end of the 19th century. This was done in reaction to the increasing prevalence of non-Euclidean geometries. A new definition of Euclidean spaces based on vector spaces and linear algebra has been shown to be comparable to the axiomatic definition. This has been shown. This definition is more often employed in modern mathematical discourse. Euclidean spaces are always made up of points, and the only way to tell them apart is to look at the characteristics that each one must have (Poole, 2014). Because every Euclidean space in a given dimension is isomorphic to every other Euclidean space in that dimension, each dimension has only one Euclidean space. As a consequence, one may operate in a specific Euclidean space, most typically the real n-space R n, which is also known as "mathbb R n" and "displaystyle R n," and is equipped with the dot product. A point's Cartesian coordinates are an n-tuple of real numbers that indicate the point's exact location in Euclidean space through an isomorphism that translates the space to the mathematical space R n mathbb R n.

Euclidean space has been used to explain the different forms that may be seen in the physical world since the time of the ancient Greeks. As a result, it has applications in a broad range of sciences, including astronomy, physics, and mechanics. The use of figures, places, and positions is also frequent in technically oriented professions such as architecture, geodesy, topography, navigation, industrial design, and technical drawing (Strang, 2022).

Outside of Euclidean geometry, Euclidean spaces may be found in a variety of fields. In the context of Euclidean geometry, vector spaces are tangent spaces of differentiable manifolds. In a broader sense, a manifold is any space that may be approached locally by spaces that obey Euclidean geometry. The overwhelming majority of non-Euclidean geometries may be represented by manifolds that are nested within larger Euclidean spaces. If desired, an ellipsoid may be utilized to provide a representation of an elliptic space. Euclidean space is often used to describe mathematical concepts that do not, by definition, contain geometry. When creating diagrams, a consistent format should be followed.

Many more forms of spaces have been studied since the late 19th century when non-Euclidean geometries were first introduced. All of these additional types of spaces are geometrically accessible in the same way that Euclidean spaces are. While they may have some qualities with Euclidean spaces, they may also have unique traits. While some of these spaces may be understood in terms of Euclidean geometry, others can be thought of as subspaces of a larger Euclidean space. The standard approach of demonstrating that a space specified by mathematical axioms is consistent with itself is to embed it inside a Euclidean space. Despite the fact that it is based on the unprovable premise that Euclidean geometry is consistent, this approach is regarded as the most dependable way for demonstrating this quality (Poole, 2014).

6.1.1.1. Affine Spaces

Metrics are used to an affine space to assess whether or not it is Euclidean. Affine spaces offer a broader range of mathematical applications. Because they may be specified over any field, they are important when seeking to obtain geometry for a number of reasons. Non-linear problems may be solved using affine spaces over complex numbers, which are an extension of Euclidean spaces. In complex affine space, a circle and a line always meet at two points. These points may or may not be completely separate from one another. As a result, complex affine spaces and affine spaces over algebraically closed fields form a large part of algebraic geometry. The study of algebraic geometries in affine spaces results in "affine algebraic varieties." The connecting link between number theory and (algebraic) geometry is given by affine spaces over algebraic number fields. In general, rational numbers provide this connection. The following assertion is illustrated: "A Fermat curve of degree greater than two has no point in the affine plane over the rationals." This is one interpretation of Fermat's Last Theorem. There are plenty others. Furthermore, the geometry of affine spaces over finite fields has received considerable attention. Elliptic curves over finite fields, for example, have a lot of applications in cryptography (Strang, 2022).

6.2. EIGEN VALUES AND EIGEN VECTORS

In linear algebra, an eigenvector is a vector that suffers a change proportionate to a constant scalar when subjected to a linear transformation. An eigenvector is also known as a characteristic vector. The Greek letter lambda is often used to represent the eigenvalue. The eigenvalue is the scaling factor for the eigenvector. A transformation matrix may modify the shape of vectors by rotating, stretching, or shearing them. When a transformation makes an eigenvector longer, it does not twist or shear; instead, it rises in the direction that corresponds to a true nonzero eigenvalue of the matrix. The amount to which an eigenvector's length rises is exactly proportional to the vector's eigenvalue. When an eigenvalue has a negative value, the orientation is flipped (Poole, 2014).

If and only if v is a nonzero vector in V and T is a linear transformation from V to itself over F, T(v) is a scalar multiple of v. T is a linear transformation if v is a nonzero vector in V. Linear transformations from an n-dimensional vector space into itself may be expressed as n-by-n square matrices, and this holds true for any vector space basis. In other words, n-dimensional vector spaces may be linearly translated into themselves. As a result, identifying eigenvalues and eigenvectors in a vector space with a finite number of dimensions using matrices or linear transformations is identical (Strang, 2022).

Eigenvalues and eigenvectors are important tools for study in the area of linear transformations. The prefix eigen-, derived from the German word eigen, which may be interpreted as "property," "characteristic," or "own," denotes "own" in this case. Eigenvalues and eigenvectors have several uses, including matrix diagonalization, matrix stability analysis, matrix vibration analysis, matrix atomic orbitals, and matrix face recognition. In the beginning, they were employed to determine the major rotation axes of rigid bodies. A linear transformation's eigenvector v may be utilized to efficiently When applied to a vector, T does not cause it to shift direction if its magnitude is larger than zero. The only thing that changes when T is applied to an eigenvector is the scalar value, which is also known as an eigenvalue. This situation might be described using a mathematical equation (Poole, 2014).

displaystyle = T (v) = T(mathbf v) Lambda'mathbf v' v

The Eigenequation is often referred to as the Eigenvalue Equation. In general, you are allowed to use whichever scalar you like. For example, if the eigenvector is zero, complex, or negative, its orientation will be inverted in the opposite way throughout the scaling process. Any point on the matrix may be freely translated in either direction along any of the eigenvectors. The Mona Lisa will serve as an instance in the context of this too basic scenario. A vector extending from the center of the painting to the location in question may be used to designate each point on the canvas. A shear mapping is the name given to the linear transformation used in this case. The points in the image's top and bottom halves are rotated to the right and left, respectively, depending on their distance from the picture's horizontal center axis. As a consequence, the vectors that represent each of the original picture points are deformed and stretched to the right or left as a result of the transformation. This modification will affect no locations along the horizontal axis. Given that the orientation is preserved during the mapping process, the eigenvectors of this transformation are any vectors that lack a vertical component and point in a straight line in a horizontal direction. Because the mapping has no effect on the length of these eigenvectors, they all have eigenvalues of one.

Given the ability of linear transformations to shift vectors into other vector spaces, eigenvectors may take on a wide variety of forms. The eigenvectors are scaled copies of the eigen functions when the linear transform is a differential operator, such as d d x tfrac d dx.

If the linear transformation matrix has n rows and n columns, then the eigenvectors may also have n rows and n columns. The solution to the eigenvalue equation may be determined by simply multiplying the two matrices together if the linear transformation is described as a matrix A with nn rows and columns (Strang, 2022).

"v" is equal to "displaystyle" in mathbf. A.

where; v is the n1 matrix's eigenvector. By first decomposing the matrix into its eigenvalues and eigenvectors, one may execute operations on the matrix, such as diagonalization (Poole, 2014).

In mathematics, ideas derived from eigenvalues and eigenvectors are often denoted by the prefix eigen-, a common mathematical prefix. The eigen system of a linear transformation is made up of the linear transformation's eigenvectors, eigenvalues, and other components. The word "eigen space" refers to the set of all T eigenvectors that have the same eigenvalue, including the zero eigenvector. The characteristic space of T associated with that eigenvalue is frequently referred to as the "eigen space." The eigen basis is a set of eigenvectors that provides structural support for a function T's domain.

It is possible to find the eigenvalues of a matrix A A by retracing the roots of each eigenvalue to the defining polynomial. This is simple with two-by-two displays, but it becomes much more complex with larger matrices.

Because the coefficients of the characteristic polynomials are just the sums of the products of the matrix's components, an exact estimate of those coefficients is achievable. It is feasible to discover all of the roots of any degree of a polynomial with the required accuracy using a variety of approaches. Rounding mistakes in coefficients are unavoidable, and in certain situations, the roots of a polynomial may be sensitive functions of the coefficients (Wilkinson's polynomial is an example of this). Even matrices constructed completely of integers become unsolvable as the sum in issue becomes larger. The determinant, often known as the constant term, is the sum of n! n! unique products for a nnn-dimensional matrix.

Accurate algebraic equations for polynomial roots are only attainable if the degree of the polynomial is four or less. The Abel-Ruffini theorem claims that it is theoretically impossible to find the roots of a polynomial with a degree of 5 or higher using a consistent, simple, or exact algebraic technique. Because each polynomial of degree n n has a corresponding matrix of order n n, the value of generality cannot be emphasized. Because there is no clear algebraic technique for determining eigenvalues and eigenvectors of matrices of rank 5 or higher, this information must be calculated numerically. It is technically difficult to arrive at an exact method for finding the roots of a polynomial of degree 3 (Poole, 2014).

6.3. ORTHOGONAL MATRICES

A unitary matrix, in reality, is the generalization of a certain kind of matrix known as an orthogonal matrix, which is always a normal matrix. Despite the fact that this chapter only deals with real matrices, the notion discussed is applicable to matrices with inputs from any field. Because dot products always yield orthogonal matrices, we may deduce that they meet the unitary condition for complex number matrices. The inner product may be better understood by analyzing a vector v in n-dimensional Euclidean space. The symbol vTv represents the square of the extent v with respect to an orthonormal basis. As a result, utilizing rotations, reflections, and their related combination operations, it is feasible to construct orthogonal matrices from linear isometries in limited dimensions. Matrix orthogonality requires the presence of orthogonal mappings. Linear algebra orthogonal transformations, on the other hand, may occur across spaces that are neither finite nor of the same dimension (Strang, 2022).

The usage of orthogonal matrices has a variety of advantages, both theoretical and practical. When the nn orthogonal matrices are multiplied together, they form the orthogonal group O(n), which is helpful in a number of applications across the mathematical and scientific fields. An O(3) group, for example, is called a molecule's point group. Floating point orthogonal matrices are vital to a variety of methods in numerical linear algebra, most notably the QR decomposition, due to the useful characteristics they give. In the MP3 compression process, a discrete cosine transform is utilized, which may be expressed as an orthogonal matrix with the proper normalization.

Numerical analysis makes use of the intrinsic qualities shared by a large number of orthogonal matrices for the purpose of numerical linear algebra. Orthogonal matrices may be useful in a number of situations, such as when constructing an orthonormal basis for a space or executing an orthogonal base change. A determinant of 1 and all eigenvalues of size 1 is a good approach to drastically boost a system's numerical stability. The multiplication of an orthogonal matrix does not increase flaws since the condition number is 1, which is the minimal minimum. Givens reflections and Householder reflections are two examples of orthogonal matrices-based methods. Both Givens and Householder reflections are forms of reflections. Because an orthonormal matrix may be inverted and obtained by simply altering the indexes, this kind of matrix should be utilized wherever feasible. Permutations are required for the functioning of several algorithms, including the widely used Gaussian elimination with partial pivoting (in which permutations do the pivoting). Because of their unique properties, matrices may be expressed in a more useful form, such as a list of n indices; yet matrices are seldom encountered in their pure form. Givens- and Householder-based algorithms often use non-standard matrix multiplication and storing techniques. A Givens rotation, for example, may reduce the complexity of a complete multiplication of order n3 to order n by altering just two rows of the matrix that is being multiplied. The extra space created by reflections and rotations has the ability to hold all of the information needed to exactly replicate the change (Poole, 2014).

6.4. PROJECTIONS

When discussing "projection" in the disciplines of linear algebra and functional analysis, it is supposed to refer to a linear transformation P P from one vector space to itself (an endomorphism). If P P always delivers the same result when applied repeatedly on the same vector, it is said to be idempotent. This is true if it always produces the same result. It has the same look as previously. In this sense, the term "projection" formalizes and expands on the concept of graphical projection. In particular, the term "projection" Examining how a projection alters the internal characteristics of a geometric object is one technique to get an understanding of the projection's influence on the object.

In certain cases, non-orthogonal projections are also known as oblique projections. Although not as prevalent as orthogonal projections, these projections are also utilized in two-dimensional drawings to depict spatial figures. They are, however, not as precise as orthogonal projections. The oblique projection is used to get the fitted value of a regression utilizing instrumental variables.

This contrasts with the orthogonal projection, which is used to derive the fitted value of a regression using ordinary least squares. The basis vectors that are used to construct them both define the range of a projection, which is the same as the kernel's complement, and the kernel itself. When the basis vectors in question are orthogonal to the kernel, we obtain an orthogonal projection. The projection will be oblique if these basis vectors are not perpendicular to the kernel; otherwise, it will simply be a projection (Strang, 2022).

6.5. MATRIX OPERATIONS

If the array or table is in the form of a matrix, a rectangular array or table of letters, numbers, or other symbols sorted in rows and columns may serve as a representation of a mathematical object or attribute. Linear algebra is simplified by matrix representations of linear transformations, which do not need any extra properties for calculation. As a result, matrices play an important role in linear algebra, and they may be utilized to describe the bulk of the properties and operations found in abstract linear algebra. Matrix multiplication, for example, may be seen as a representation of the process of building linear maps (Figure 6.2).

$$A \cdot B = \begin{bmatrix} R_1C_1 & R_1C_2 \\ R_2C_1 & R_2C_2 \end{bmatrix} = \begin{bmatrix} 3\cdot 1 + 4\cdot 0 + 1\cdot 8 & 3\cdot 5 + 4\cdot 9 + 1\cdot 7 \\ 6\cdot 1 + 2\cdot 0 + 5\cdot 8 & 6\cdot 5 + 2\cdot 9 + 5\cdot 7 \end{bmatrix}$$

$$= \begin{bmatrix} 3 + 0 + 8 & 15 + 36 + 7 \\ 6 + 0 + 40 & 30 + 18 + 35 \end{bmatrix} = \begin{bmatrix} 11 & 58 \\ 46 & 83 \end{bmatrix}$$

Figure 6.2. Matrix multiplication color illustration.

Source: Speakman & Courtney (2021); Scientific Figure on ResearchGate.

There is not always a one-to-one correlation between matrices and linear algebra. This is true in graph theory for both incidence and adjacency matrices. Unless otherwise noted, all matrices relevant to linear algebra that are treated in this chapter may be understood in terms of linear maps or are represented by linear maps. The study of matrices is heavily reliant on square matrices, which have the same number of rows and columns. One of the most common is built from non-commutative rings of a certain size, which serve as the essential construction blocks. If the determinant of a square matrix is not equal to zero and the matrix's eigenvalues correspond to the roots of a polynomial, the matrix is said to be invertible. One of the most important numbers to grasp about a square matrix is its determinant (Poole, 2014).

Transformations (such as rotations) of coordinates and other geometric objects are often expressed in geometry using matrices. Translations and reflections are two examples of these changes. Many of the computational challenges that must be tackled in numerical analysis are approached by doing matrix computations. These calculations may sometimes require dealing with exceedingly big matrices. Matrix algebra may be found in almost every branch of mathematics and science, including geometry and numerical analysis, either directly or indirectly. Matrix theory is a branch of mathematics that investigates matrices as objects of study. Beginning as a subfield of linear algebra, the discipline quickly grew to embrace concepts and principles from other subfields and areas of study, including graph theory, algebra, combinatorics, and statistics. There are several ways that may be used to address matrix issues. It is feasible to solve a broad range of problems using either iterative or direct techniques. When attempting to discover the eigenvectors of a square matrix, for example, we look for a series of vectors represented by xn that converges to a single eigenvector as the value of n approaches infinity (Strang, 2022).

In order to choose the best technique for accomplishing each job, the efficacy and accuracy of each algorithm must be evaluated. These are the subjects covered by the branch of study known as numerical linear algebra. The complexity and stability of the algorithm are always important elements in any circumstance involving numbers. The number of basic operations, like as additions and scalar multiplications, needed to perform a certain algorithm, such as matrix multiplication, is used to define complexity. Estimates or upper constraints on the number of operations may be computed. According to the definition above, multiplying two n-by-n matrices together requires n3 multiplications, one for each of the n2 components created by the multiplication. The Strassen approach, which needs only n2,807 times the amount of multiplications, outperforms this "naive" method. A more involved technique employs a bigger number of computer hardware components. A wide range of real-world applications may give extra information about matrices. Sparse matrices, which include a high number of negative elements, are a good example of this. The conjugate gradient technique, for example, is one strategy that may be used to solve sparse matrices Ax = b linear equations (Poole, 2014).

When an algorithm produces consistent outputs across a wide range of input values, we say that the algorithm is numerically stable. Matrix norms might involve the conditioning of linear algebraic problems, such as the identification of inverse matrices. Although arrays are included in practically all programming languages, matrix instructions are often missing. Instead, the majority of popular programming languages handle matrix operations on arrays through external libraries that are widely accessible to the general public. The manipulation of matrices was one of the early uses of numerical computation. Dartmouth BASIC's second edition, released in 1964, included matrix arithmetic on arrays as a new feature. Several technically capable desktop computers were equipped with ROM cartridges with matrix-specific BASIC instructions as early as the 1970s. Computing matrices may be made simpler with the help of a variety of mathematical programs and computer languages, such as APL, which were built expressly for dealing with matrices (Strang, 2022).

There are several approaches to simplifying the process of displaying matrices. These methods are also known as matrix factorization and matrix decomposition, among other names. All of these methods have applications, whether it is because specific matrix operations are algorithmically simpler to perform for specific types of matrices, or because they retain matrix attributes such as the determinant, rank, and inverse, allowing these values to be determined after the transformation has been performed.

The LU decomposition factor matrices are obtained by multiplying the lower triangle (L) and upper triangle (U) matrices together. After determining this decomposition, the forward and reverse substitution approach may be used to the linear system to provide a more precise solution. Triangular matrices' inverses are also easier to calculate algorithmically than other forms of matrices. The Gaussian elimination method may be used to convert any matrix to row echelon form for easier analysis. The matrix is always multiplied by basic matrices, which either reverse the order of the rows or columns or combine multiples of a single row with another row. A matrix A may be written as the product UDV using the singular value decomposition technique, where U and V are unitary matrices and D is a diagonal matrix. This is one method of creating a matrix (Poole, 2014).

The matrix VDV1 is the product of A's eigen decomposition, also known as its diagonalization, if D is a diagonal matrix and V is an admissible invertible matrix. We argue that this form is only diagonalizable if and only if it can be utilized to represent A. The Jordan decomposition, as shown on the right, converts a matrix into the Jordan normal form, which can be applied to any matrix in general and produces matrices with no nonzero entries other than the eigenvalues 1 through n of A located on the main diagonal and possibly entries equal to one directly above the main diagonal. On the right is the Jordan decomposition. The eigen decomposition may be utilized in a variety of ways, including calculating the nth power of A and performing n-fold iterative matrix multiplication. Finding the power of a diagonal matrix by exponentiating A is therefore a significantly more difficult process than calculating the suitable powers for the diagonal members. The matrix exponential eA is often used in linear differential equations, matrix logarithms, and matrix square roots. It is feasible to avoid mathematically ill-conditioned situations by using extra approaches such as the Schur decomposition (Strang, 2022).

6.6. LINEAR DEPENDENCE

A group of vectors is said to be linearly independent in vector space theory if there is no nontrivial linear combination of the vectors that equals the zero vector. This requirement must be satisfied for a linear combination of vectors to be considered nontrivial. If there is a combination that meets this requirement, the vectors are said to be linearly dependent on one another. Dimensions may be defined by referring to these ideas. The highest number of linearly independent vectors may be used to assess if a vector space's dimensions are finite or infinite. A good grasp of linear dependency, as well as a mechanism for detecting whether or not a collection of vectors in the space is linearly dependent, are required for quantifying a vector space (Poole, 2014).

"It is 3 miles to the north of here and 4 miles to the east of here," a statement like this would say. Because geographic coordinates may be understood as a two-dimensional vector space, this approach is appropriate for finding the place (ignoring height and surface curvature). Following that, they may add something like, "The location is 5 miles to the northeast of here." The concluding statement is correct; nonetheless, the geographical data offered were superfluous. In this example, the coordinates "3 miles north" and "4 miles east" are not linearly dependent on one another. To put it another way, neither the east vector nor the north vector can offer an accurate description of the other. Because the third vector is a linear combination of the first two, just two of the three "5 miles northeast" vectors are necessary to identify a particular place on a plane. This is due to the vector set being linearly dependent, which means that the third vector is reliant on the other two vectors. Furthermore, if the height is taken into account, a third vector is added to the collection of vectors that are linearly independent. Each point in n-space must be represented by an n-vector group that is linearly independent of one another. The idea of linear independence is a fundamental building element in linear algebra. If there is no linear combination that can be constructed between two or more vectors, they are termed linearly independent. If, on the other hand, one may be represented as a linear combination of the others, then their relationship is one of linear dependence (Strang, 2022).

6.7. MODULE THEORY

The axioms that define a vector space do not account for the presence of multiplicative inverses in the fields under consideration. A module over R, often known as an R-module, is the structure formed when a ring R is substituted for the field of scalars. The primary distinction between the definitions of linear independence, span, basis, and linear mappings (also known as module homomorphisms) for modules and vector spaces is that modules may exist without a basis if R is not a field. This is the primary difference between the two sorts of definitions. In contrast to finitely generated modules, free modules have no underlying structure. An infinite collection spans finitely created modules. Matrix representations may be used to define module homomorphisms between freely constructed modules of a limited size. A square matrix over a commutative ring may be inverted only if the matrix's determinant has a multiplicative inverse in the ring. In field theory, determinants can only exist if the ring on which they are specified is commutative (Poole, 2014).

A vector space's dimension may only be given in one method, and that is by isomorphism. Even if one limits oneself to modules that are formed in small numbers, the fact remains that there is no such thing as a thorough categorization of modules in general. The homomorphism of free modules, on the other hand, acts as the cokernel for each and every module. Because of the equivalence between periodic addition and multiplication by an integer, Abelian groups have the potential to be employed for identifying modules over integers. The use of modules on the basic ideal domain may generalize a major chunk of abelian group theory. Because every submodule of a free module is free over a primary ideal domain, the basic theorem of finitely generated abelian groups may be applied to modules over a primary ring with ease. This provides more flexibility in the study of abelian groupings (Strang, 2022).

Algorithms may be used to solve linear equations and systems of linear equations for a broad range of rings. However, the computational cost of these techniques is often much greater than that of other algorithms in the same sector (Poole, 2014).

6.8. MULTILINEAR ALGEBRA AND TENSORS

Multilinear algebra is concerned with linear translations and transformations with a large number of variables. The topic of this inquiry lends itself to the development of the idea of dual space.

Dual space is the vector space V* generated by linear mappings f: V F, where F is the scalar field. The phrase "dual space" refers to this dual space architecture. It is feasible to build multilinear mappings with the notation T: Vn F using the tensor product of the component tensors of V*.

A vector space is called an algebra if it allows the solution of the bilinear vector product V V V in addition to vector addition and scalar multiplication. In mathematics, scalar multiplication and vector addition are both considered basic operations. The algebra of square matrices and the algebra of polynomials are two examples of associative algebras with associated vector products. Both of these algebras are concerned with polynomials.

6.9. TOPOLOGICAL VECTOR SPACES

In order to make non-finite dimension vector spaces viable, extra structure is typically required. A norm function governs the "size" of the elements in a normed vector space, which, along with the vector space, constitutes the normed vector space. A normed vector space is sometimes known as a normed vector space. The norm provides a topology for constructing continuous maps as well as a metric for analyzing component part separation. A Banach space is a metric space that is totally complete. Limits and completeness may also be specified using this measuring tool.

Hilbert spaces are a kind of Banach space characterized by the existence of metric inner products in the form of conjugate symmetric sesquilinear forms. The Lp spaces, which are Banach spaces, and in particular the L2 space of square integrable functions, which is the sole Hilbert space among them, are the primary foci of functional analysis study. Linear algebra and other kinds of mathematical analysis are utilized in functional analysis to conduct analysis on a variety of function spaces. Many areas may benefit from functional analysis, including quantum physics, partial differential equation theory, digital signal processing, and electrical engineering. Furthermore, it gives a theoretical framework for the Fourier transform and the operations associated with it (Poole, 2014).

6.10. APPLICATIONS OF LINEAR ALGEBRA IN MATHEMATICS

The ability to apply linear algebra to a broad range of various mathematical subfields makes it an indispensable tool for any scientific subfield that sets a high value on mathematical analysis. There are many broad categories that may be used to define these software programs. Every interior space model must have a strong geometric base. Geometry is important in many branches of study that delve at this portion of the world. Perspective, computer vision, and computer graphics may be used to explain the relationship between a scene and its planar representation; mechanics and robotics can be used to describe the movements of rigid bodies; geodesy can be used to describe the shape of the earth, and so on. Coordinate calculations are necessary to conduct research into particular cases; nonetheless, synthetic geometry is widely employed in all of these applications due to its capacity to give broad descriptions and a qualitative approach. This issue will need a large amount of linear algebra (Strang, 2022).

The study of function spaces is what is meant by "functional analysis." These vector spaces have been enhanced by adding structure in the same way as Hilbert spaces have been improved. For applications of functional analysis, such as those in quantum physics (wave functions), knowledge of linear algebra is essential. The bulk of events in the physical world are explained using partial differential equations. When seeking to find answers, it is standard practice to divide the problem region into a large number of extremely tiny cells that are continually connecting with one another and exchanging information. Because of this relationship, working with linear systems involves the employment of linear functions. When working with nonlinear systems, it is standard practice to represent the kind of interaction illustrated here using linear functions. [b] In both cases, the matrices are often of significant size. The Earth's atmosphere is often shown as a grid with cells 100 kilometers wide and 100 meters tall. The bulk of individuals who try to predict the weather use this technique (Poole, 2014).

Almost all scientific computations need knowledge of linear algebra. As a consequence, the development of linear algebra techniques has advanced significantly. The most often utilized programming languages are BLAS and LAPACK. Some of them improve system performance by automatically adjusting the algorithms while the application is executing. This ensures that everything runs as smoothly as possible. This is accomplished by optimizing the algorithms for the computer's particular hardware configuration (such as the size of the cache and the number of processor cores), allowing the computer to do its duties more rapidly. Certain kinds of central processing units, particularly GPUs, utilize matrix structures to accelerate linear algebraic calculations (Strang, 2022).

Encryption and decryption are being studied because these procedures are utilized to transport sensitive information. If one so desire, strong encryption technologies may be utilized in online business and communication. Modular arithmetic is required to encode and decode messages using these techniques. Matrix transformation is a notion that might be employed in the creation of more easy encoding systems. Here is another use of linear algebra, which is a field of mathematics that counts options. Throughout the game, players will be expected to devise tactics for going ahead with all of these many alternatives. Psychologists employ the social interaction hypothesis while doing research that explores how one player's actions connect to those of other players.

Although the most popular applications of game theory include card games, board games, and other types of competitive play, the theory may also be used to explain how armies fight in a war. Signals like as audio and video are encoded and manipulated using linear algebra throughout the signal processing process. It is also required for this kind of signal analysis. Finding the optimum linear function may be performed via the use of linear programming, an optimization approach. In the realm of computer science, data analysts employ a wide range of linear algebraic methodologies to solve challenging issues. Algorithm-based prediction-making algorithms Prediction algorithms employ basic principles from linear algebra to build linear models (Poole, 2014).

Linear transformations need knowledge of planes and lines, mapping, and vector spaces, which form the foundation of linear algebra. As a result, understanding the many applications of linear algebra is critical. Linear algebra is often referred to as a mathematical area, although the phrase "the mathematics of data" better reflects its significance. Matrixes and vectors are used to express the informative language. Linear algebra is concerned with linear combinations. Mathematical operations on a matrix or vector's rows may be used to generate new columns of numbers or arrays.

CHAPTER 7

Mathematical Graphs

Contents

7.1. INTRODUCTION

G = (V, E), where V denotes a set of paired vertices with edges (or, alternately, links or lines) and E denotes a set of vertices. This section defines the terms graph, directed graph, and multigraph.

Endpoints (x and y) of an edge are represented by vertices (x and y). The edge is said to be incident on x and y and to link x and y. If a vertex does not belong to an edge, it cannot be linked to another vertex. In a multigraph, many edges may share the same pair of endpoints. Multigraphs are commonly referred to as graphs in formal writing (Kuchment, 2013) (Figure 7.1).

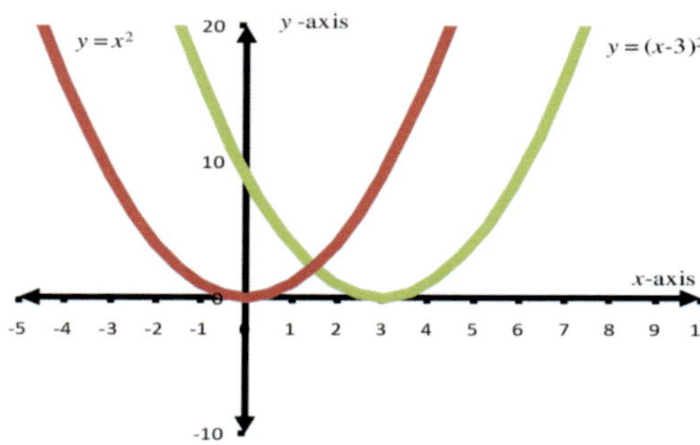

Figure 7.1. Graph showing the translation of $y = (x-3)^2$.

Source: Benning & Agyei (2016); Scientific Figure on ResearchGate.

Loops, which are edges that link two vertices, are uncommon in graphs. Vertices must be able to share the same node once more for E to allow cycles. Graphs containing loops or evident loops are examples of generalized graphs.

Because it is widely assumed that the number of vertices V is restricted, the number of edges must likewise be constrained. Because the great majority of discoveries concerning finite graphs do not apply to infinite networks or necessitate other techniques of reasoning, infinite graphs are frequently regarded as a specific sort of binary connection, even though they are rarely investigated (Lengyel, 2011).

A graph is said to be vacant if its set of vertices, and hence its set of edges, is empty. The vertex count, or |V|, determines a graph's rank. |E| denotes the number of edges in a graph or its extent. Even though the magnitude of a non-empty network might be |V| plus |E| at times, when analyzing an algorithm's computational complexity, it may occasionally be 0. The number of edges that contact a vertex determines its degree or valency; in graphs, an edge is enumerated twice for each loop (Kuchment, 2013).

A graph of order n has n (n 1)/2 edge degrees and a maximum degree of n 1 (or n + 1 if loops are allowed, as loops raise the degree by 2).

The edges that link a graph's vertices have a symmetrical adjacency relationship. If x, y is an edge, then x and y are close together. A graph is entirely described by the adjacency matrix A, an n-square matrix that indicates the number of linkages from vertex I to vertex j. Because edges in a simple graph cannot begin and end at the same vertex, Aij must be either 0 (disconnection) or 1 (connection). Aii must be negative as well. For multigraphs (graphs with numerous edges linking vertices), some or all of Aij will be positive integers, but some or all of Aii will be positive integers for self-looping graphs. Undirected graphs' adjacency matrix will be symmetric (i.e., Aij = Aji).

7.2. CARTESIAN GRAPHS

A Cartesian coordinate system on a plane is a type of coordinate system used in geometry that identifies each point with a pair of real numbers known as coordinates. Numerous individuals employ the phrases "coordinate lines," "coordinate axes," and "axes" (plural of axis). At the origin, which has the coordinates (0, 0), they all converge. The three Cartesian coordinates, which are the signed distances between a point and three mutually perpendicular planes, can also be used to determine the three-dimensional position of each point. A formal representation of a point in an n-dimensional Euclidean space is its n-Cartesian coordinates. There are furnished signed distances between the point and n fixed, perpendicular hyperplanes (Lengyel, 2011).

The revolutionary 17th-century Cartesian coordinate system of René Descartes enabled the first systematic connection between geometry and algebra. Geometric shapes (such as curves) can be represented by the Cartesian coordinate system using equations based on the points' coordinates. A circle of radius 2 with its center at the origin of the plane is, for instance, the set of all points whose x and y coordinates satisfy the equation x2 + y2 = 4 (Kuchment, 2013).

Analytical geometry, which is based on Cartesian coordinates, provides geometric interpretations for a variety of other mathematical disciplines, including group theory, multivariable calculus, differential geometry, complex analysis, and linear algebra. The vast majority of practical applications of geometry, including those in physics, engineering, and astronomy, also employ cartesian coordinates. A function graph is an excellent illustration of this concept. They are the most common coordinate system used in computer graphics, computer-assisted geometric design, and other data processing applications requiring geometry (Lengyel, 2011) (Figure 7.2).

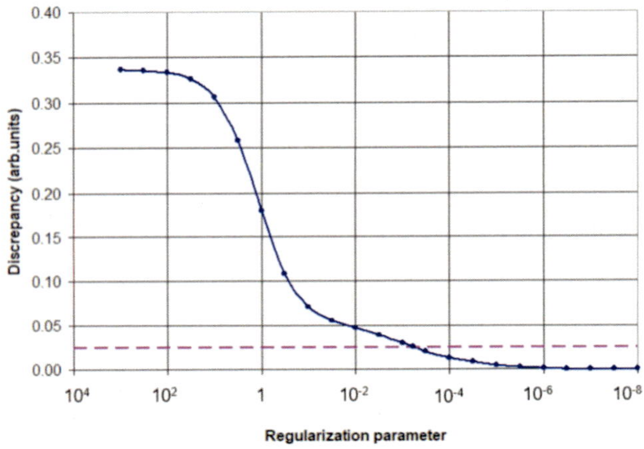

Figure 7.2. Dependence of the discrepancy between the experimental and calculated frequencies on the value of the regularization parameter α in the process of optimization.

Source: Kochikov, Stepanova, & Kuramshina (2022).

René Descartes, a French mathematician, and philosopher, coined the term "Cartesian" in the Netherlands in 1637. Pierre de Fermat, who independently conducted three-dimensional research, independently made the discovery but chose not to make it public. Nicole Oresme, a French theologian, utilized coordinate systems long before Descartes and Fermat created them. Both Descartes and Fermat examined a single axis and evaluated a variable length about it. Frans van Schooten and his pupils proposed the use of two axes in 1649, following the Latin translation of Descartes' La Géométrie. These remarks offered a multitude of original ideas while attempting to clarify Descartes' work's principles. The introduction of the Cartesian coordinate system was crucial to Isaac Newton and Gottfried Wilhelm Leibniz's development of calculus. The two-coordinate vector space of the plane was later expanded into a broader concept of vector spaces. Since Descartes, several additional coordinate systems have been developed, including spherical and cylindrical coordinates for three-dimensional space and polar coordinates for the plane (Kuchment, 2013).

7.2.1. One Dimension

Before a Cartesian coordinate system can be used to characterize a one-dimensional space or a straight line, the origin, length unit, and orientation of the line must be determined. When an orientation identifies which of the two half-lines supplied by O is the positive half and which is the negative half, the line "is oriented" (or "points") from the negative half to the positive half. Consequently, the distance between each point P on the line and O can be determined contingent on whether the half-line incorporates P. A number line is a linear line that corresponds to a particular Cartesian coordinate system. Each real number occupies a separate position on the line. On the other hand, each point on the line can be viewed as a number in a series of ordered numbers, such as real numbers (Lengyel, 2011).

A number line is a graphic of a graded straight line used to depict real numbers in elementary mathematics. It is assumed that each point on a number line corresponds to a real number, and vice versa. On a line, numbers are typically represented by distinct, equally spaced marks. Although it appears to display only the integers –3 to 3 and the numbers in between, the line contains all real numbers and extends indefinitely in both directions. This is a common method for teaching elementary addition and subtraction, particularly when working with negative numbers. In advanced mathematics, the real line, also known as the real number line or the real line, is the set R of all real numbers. It is demonstrated that the object is a geometric space, or more specifically, a one-dimensional real coordinate space or a one-dimensional Euclidean space containing the Euclidean line. It is also referred to as topological space, measure space, vector space (or affine space), and linear continuum. As with the set of real numbers, the letter R or the word "R" in bold on the whiteboard is commonly used to denote the real line. It is commonly referred to as R1 or E1 to highlight its significance as the first real or Euclidean space (Kuchment, 2013).

John Wallis first proposes the number line as an instrument for operations in his Treatise on Algebra. Using the metaphor of strolling, Wallis depicts addition and subtraction on a number line as forward and backward motion. The numbers 1 through 12 are arranged from left to right in an older version of John Napier's explanation of the marvelous table of logarithms; however, the operations are not specified. Contrary to prevalent belief, the original

La Géométrie by René Descartes utilized a coordinate system rather than a number line. The entirety of Descartes's work, for instance, consists of abstract sums rather than lines conveying actual values.

Even though a number line is typically depicted as horizontal, the vertical axis (y-axis) of a Cartesian coordinate plane also functions as one. Both extremities of the line are marked with arrowheads to signify that it can be drawn indefinitely in both positive and negative orientations. Negative numbers always follow zero, while positive numbers always follow it. Alternatively, the orientation of the numerals may be indicated by a single arrowhead. According to geometry principles, a line with no endpoints is an infinite line, a line with one endpoint is a ray and a line with two endpoints is a line segment. The line is infinite in both positive and negative orientations.

On the number line, the first number is larger if it is located to the right of the second number, or vice versa. The fact that the first number is less than the second number or that the absolute value of the second number is less than the first number demonstrates their difference. Subtraction is the operation utilized to determine the difference.

Thus, the length of a line segment connecting 0 to another integer can be used to represent the magnitude of the latter number. Two numbers can be added by "picking up" the length from 0 to one of the numbers and "setting it down" on top of the other number.

Consider the two figures below: Place a length between 0 and 5 to the right of 5, followed by another length between 0 and 5, so that 5 multiplied by 3 equals 5 plus 5 plus 5, and 5 multiplied by 5 equals 5 plus 5 plus 5. Since the method ends at 15, we can conclude that 5 3 = 15. As a consequence, there are a total of three lengths (Lengyel, 2011). In the example below, the division is performed as follows: To divide 6 by 2 or to determine how many times 2 can fit within 6, move the previous length to the right of its original position, with the end that was previously at 0 now at 2, and then move the length to the right of its most recent position. Therefore, the precise extent between 0 and 6 is 2. Because 6 2 = 3, six were divided into three lengths of two, indicating that three instances of two may exist within six (Kuchment, 2013). On a number line, an interval is the space between two numbers. The interval is considered closed if both numbers are present; otherwise, it is considered open. When only one of the integers is present, it is said that the interval is half-open. Rays are all the points that radiate from a singular point indefinitely. The ray is closed if the designated point is included; otherwise, it is open.

7.2.2. Two Dimensions

In two dimensions, a pair of Cartesian coordinates consists of an ordered pair of perpendicular lines (axes), a single unit of length for each axis, and an orientation for each axis. A rectangular or orthogonal coordinate system is another name for it. Each axis becomes a number line when the origins of both axes are near their junction. A random point P is chosen as the junction point of a line perpendicular to each axis, and it is then assigned a numerical value. The order of the numbers represents P's Cartesian coordinates. Because of its inverted structure, the coordinates of point P may be utilized to find it (Lengyel, 2011).

The coordinate system's origin is the point where the two axes cross and the first and second coordinates are the abscissa and ordinate of point P, respectively. Coordinates are frequently stated as two integers separated by commas and enclosed by parentheses, such as (3, 10.5). The positive half-axes' one-unit nodes are thus positioned at (1, 0) and (0, 1), respectively, while the origin is at (0, 0).

In mathematics, physics, and engineering, the first axis is commonly shown as horizontal and pointing to the right, whereas the second axis is vertical and pointing upwards. However, in certain computer graphics systems, the ordinate axis may point downward. However, the two coordinates are frequently represented as X and Y, or x and y. The axes are hence known as the X- and Y-axes, and legend has it that the last group of letters in the alphabet symbolize indeterminate numbers. The first letter of the alphabet relates to fixed values (Kuchment, 2013).

A Cartesian plane is a Euclidean plane with a Cartesian coordinate system. The Cartesian plane is home to canonical representations of the unit circle, whose radius equals the length unit and whose center is at the origin, the unit square, whose diagonal ends at (0, 0), the unit hyperbola, and a variety of other geometric objects. The plane is divided into four quadrants by the two axes. In the first quadrant, all of the coordinates are positive. There are several ways to identify and number the quadrants. When the coordinates of a point are (x, y), the distances from the X and Y axes are $|y|$ and $|x|$, where $|\,|$ represents the absolute value of an integer.

7.2.3. Three Dimensions

A three-dimensional space is represented by a Cartesian coordinate system as a collection of three ordered perpendicular lines (the axes) that all cross at the same place (the origin). Each axis is parallel and has the same length. Each axis in three dimensions, like in two dimensions, forms a number line. Consider a hyperplane perpendicular to each coordinate axis that passes through each point P in space and assign a numerical value to the point where the hyperplane and the axis cross. In the stated sequence, these three values indicate P's Cartesian coordinates. The point P is found by using its three coordinates and the reverse construction. The distance between point P and the hyperplane produced by the other two axes may also be computed using each of its coordinates, with the direction determined by the sign of the associated axis. Each pair of axes forms a coordinate hyperplane. These hyperplanes divide the universe's eight octants (Lengyel, 2011).

Three integers (or algebraic equations) are typically wrapped in parentheses and separated by commas when expressing coordinates, as in (3, 2.5, 1) or (t, u + v,/2). As a result, the origin is at (0, 0), and the unit points on the three axes are (1, 0, 0), (0, 1, 0), and (0, 0).Although the terms abscissa, ordinate, and applicate are commonly used, there are no formal names for the coordinates on the three axes. Coordinates are commonly represented by the letters X, Y, and Z, or x, y, and z. As a result, the letters X, Y, and Z stand in for the axes. The three coordinate hyperplanes are the XY-plane, the YZ-plane, and the XZ-plane, in that order (Kuchment, 2013).The first two axes are commonly expressed or drawn as horizontal in mathematics, physics, and engineering, whereas the third axis points upward. In this case, the third coordinate might be height or altitude. The right-hand rule is a method in which the first and second axes are aligned so that their 90° angle looks anticlockwise when seen from the origin (0, 0, 0).

7.3. DIRECTED GRAPHS (DIGRAPHS)

In mathematics, a directed graph (or digraph) consists of vertices connected by directed edges, also known as arcs. A directed graph cannot have multiple arrows that all point in the same direction and share the same source and target nodes, per the above definition. Other authors, however, have proposed a broader definition that permits directed graphs to have a large number of arcs (by permitting the arc set to be a multiset). It is common to refer to these as directed multigraphs (or multidigraphs). However, whereas the antecedent definition permits loops in directed graphs (arcs that connect nodes directly), some authors have proposed a stricter definition that prohibits loops in directed graphs. Loop digraphs are directed graphs that contain loops, whereas simple directed graphs do not.

A directed graph, also known as a digraph or D, consists of vertices V (D), edges E (D), and a function that assigns a pair of ordered vertices (u, v) to each edge e. We refer to u as "e's tail," v as "e's head," and v as "e's ends." If an edge has both a tail (u) and a head (v), we can use the notation (u, v) to denote that it extends from u to v. The ordered pair (V (D), E(D) can be used to more simply express a digraph D. If u = v, then the edge e = (u, v) of the digraph D is a loop. If the tails and leaders of two edges e, f match, then they are parallel. If D lacks loops and parallel edges, it is deemed straightforward. As with undirected graphs, it is beneficial to depict directed graphs with illustrations that depict each vertex as a single point and each edge from u to v as a curve directed from point u to point v (this direction is typically denoted by an arrowhead) (Lengyel, 2011).

7.4. WEIGHTED GRAPHS

A weighted graph is one in which each edge is assigned a weight, a numeric value. A network consisting of five nodes and seven edges, each of which carries its weight. For instance, the edge weight of 0,3,0,3 and 1,2,1,2 is 11. Weighted graphs are commonly used to model real-world objects and node-to-node relationships. The nodes in the preceding example are cities, and the edges are routes, rendering it a component of Google Maps. The duration/distance between two cities determines the weight of each edge. The aforementioned graph is a weighted undirected graph with undirected edges. In addition, directed weighted graphs are feasible. Rather than an adjacency list, an adjacency matrix can be used to represent an unweighted graph. One of these topologies may be used to symbolize a weighted graph (Kuchment, 2013).

7.5. COMPLETE GRAPHS

In graph theory, a complete graph is a basic, undirected graph in which each pair of different vertices is connected by a single edge. A full digraph is a directed graph with two different edges (one in each direction) linking each pair of distinct vertices. The work of Leonhard Euler on the Seven Bridges of Königsberg in 1736 is commonly regarded as the beginning of graph theory. Nonetheless, Ramon Llull's 13th-century book has pictures of entire graphs with vertices placed on regular polygon points. This representation is sometimes referred to as a "mystical rose." The full n-vertex network is represented by Kn. The German name for a full graph, on the other hand, is vollstandiger Graph, which is lacking the letter K. According to some sources, the notation honors Kazimierz Kuratowski's contributions to graph theory. According to some, the letter K in this notation indicates the German word 'vollständig.' Kn is a regular graph with degree n – 1 and n (n – 1)/2 triangular edges. Every complete graph is a maximum clique in and of itself. They are all maximally linked since the collection of all vertices is the only vertex cut that can disconnect the graph. An empty graph is the inverse of a full graph. When all of the edges in a complete graph are assigned as orientations, the resultant directed graph is a tournament. Kn is divided into n separate trees, each with i vertices. Ringel's conjecture investigates whether the complete graph K2n+1 is divided into n-edged tree copies. For sufficiently big n, this is known to be true. A (n – 1)-simplex's edges can be represented as a graph with n vertices. In geometry, K4 is a tetrahedron, K3 is the edge set of a triangle, and so on. The whole graph K7 is the skeleton of the Császár polyhedron, a nonconvex polyhedron with torus topology. Each neighboring four-dimensional polytope has a full skeleton (Lengyel, 2011).

All planar graphs K1–K4 are present. Kuratowski's theorem indicates that a graph is only deemed planar if it lacks K5 or the entire bipartite graph K3,3, and Wagner's theorem shows that the same result applies when graph minors are used instead of subdivisions. Every planar representation of a network with five or more vertices, on the other hand, must include a crossover. K6 serves a similar purpose as one of the minors barred from linkless embedding as a member of the Petersen family. In other words, every K6 embedding in three-dimensional space, as established by Conway and Gordon, has at least one pair of linked triangles. Conway and Gordon also established that a nontrivial knot-like Hamiltonian cycle exists in every three-dimensional embedding of K7 (Kuchment, 2013).

7.6. BIPARTITE GRAPHS

When depicting interactions between two separate groups of items, bipartite graphs generally emerge spontaneously. A graph displaying football players and organizations is a great example of an affiliation network, a type of bipartite graph that is commonly used in social network research. There is a link between athletes and companies who have previously collaborated. Bipartite graphs are easily created in the (NP-complete) railway optimization problem, where the goal is to locate a finite collection of train stations where each train visits at least one of the specified stations. The input for the task is a train timetable with stops. This problem may be represented as a bipartite graph with a vertex and an edge for each train and station combination as a dominating set problem. The third image comes from the world of numismatics. An antique coin's obverse and reverse are two good components of the design. Numismatists portray the development of new currencies using bipartite graphs. Knig's Theorem states that the size of the greatest matching in a bipartite graph is equal to the size of the smallest vertex cover. The number of vertices is equal to the size of the biggest independent set plus the size of the largest matching, according to a variation of this theorem. Any graph with no isolated vertices has as many vertices as the sum of the minimal edge cover and maximum matching. The size of the minimal edge cover in bipartite graphs corresponds to the size of the biggest independent set, and the total of the minimal edge cover and minimum vertex cover sizes equals the number of vertices. When Knig's theorem is combined with this equality, the following results are obtained (Lengyel, 2011). According to a second set of identical findings, all bipartite graphs, complements, corresponding line graphs, and complements of corresponding line graphs are perfect. This is a variation of Knig's thesis. Because bipartite graph complements have a chromatic number of two and a maximum clique size of two, they are more difficult to show. One of the discoveries that led to the notion of perfect graphs was this. Another interpretation of Knig's theorem holds that the line graph complements of perfect graphs are also perfect and that any bipartite graph has edge coloring with the same number of colors as its maximal degree (Kuchment, 2013) (Figure 7.3).

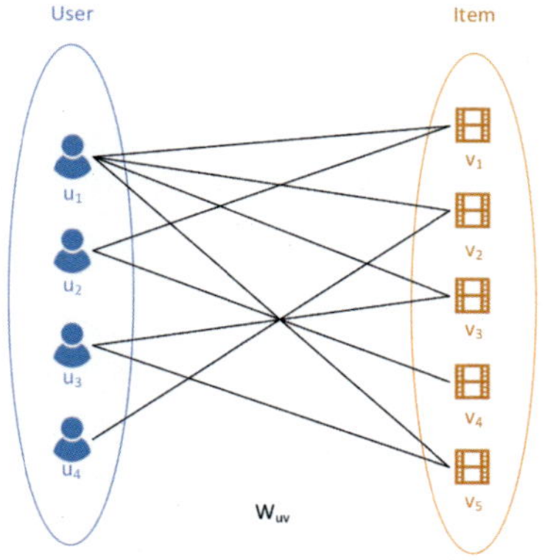

Figure 7.3. An example of bipartite graph $G = (U, V, E)$, which has two types of entities U and V.

Source: Ma, Dong, Wang, Li, & Sun (2020).

A graph is bipartite if it has no odd cycles as subgraphs, and perfect if it has no odd cycles or their counterparts as induced subgraphs. This assumption is supported by the strong perfect graph theorem. Bipartite graphs, line graphs of bipartite graphs, and their complements are four of the five fundamental kinds of perfect graphs used in the proof of the strong perfect graph theorem.

In linear time, the depth-first search technique may identify whether a graph is bipartite and produce a two-coloring or an odd cycle if it is. The primary goal is to traverse the depth-first search forest and assign a different color to each vertex in the depth-first search forest than its parent. As a result, the edges connecting vertices to their progenitors will be two-hued; nevertheless, certain non-forest edges may be wrongly colored. Each non-forest edge in a depth-first search forest has two endpoints, one of which is an ancestor of the other. When a depth-first search finds one of these non-forest edges, it should examine if the two vertices have separate colors, because each non-forest edge has two extremities, one of which is the ancestor of the other. In addition, the approach shows the irregular cycle generated by the miscolored edge, the forest path from ancestor to descendant, and the conclusion that the graph is not bipartite. If the algorithm is unable to locate an odd cycle of this type, each edge must be correctly colored (Kuchment, 2013).

Breadth-first search, like depth-first search, may be used to replace depth-first search. Each node in the search forest is assigned a color that contrasts with that of its progenitor in breadth-first order. An irregular cycle is generated when a colored vertex is connected to another colored vertex of the same color via an edge and the pathways in the breadth-first search forest that connect its two endpoints to their lowest common ancestor. This approach must have found enough coloring to infer that the graph is bipartite and there is no irregular cycle.

Bipartite networks are widely used in modern coding theory, notably for decrypting incoming codewords. Tanner and factor diagrams are two examples. Tanner graphs are bipartite graphs with codeword digits as vertices on one side and sum-to-zero codeword digit combinations on the other. For the probabilistic decoding of LDPC and turbo codes, a factor graph, which is very similar to a belief network, is used (Lengyel, 2011).

Petri nets are used in computer science as a mathematical modeling and simulation tool for concurrent systems. A system is represented as a bipartite directed graph with two types of nodes: "place" nodes, which keep resources, and "event" nodes, which generate and/or consume resources. The nodes and edges of the system are similarly of little use. Petri nets make use of bipartite-directed graphs and other features to make system simulations and mathematical behavior descriptions easier.

To describe where points and lines occur within configurations, projective geometry uses bipartite graphs known as Levi graphs. Cycles with a length of four are incompatible with Levi graphs, hence their diameter must be at least six. This is because Levi graphs have a geometric feature in which each pair of lines crosses only once, and each pair of points is connected by a single line (Kuchment, 2013).

7.7. PLANAR GRAPHS

You can construct a planar graph on a plane by enclosing it within the plane or by overlapping only its extremities. As a consequence, it can be drawn without any boundaries overlapping. An example is a plane graph, also known as a planar embedding of the graph. A plane graph is a planar network in which each node corresponds to a plane point and each edge corresponds to a plane curve on that plane, except for their extreme points, all curves being connected. The farthest locations of each curve are points mapped from its end nodes (Lengyel, 2011).

Using stereographic projection, every graph that can be drawn on a plane can also be drawn on a sphere, and vice versa. Both rotation systems and combinatorial mappings can be used to encode planar graphs. A planar map is a compilation of topologically comparable drawings on the sphere constrained by additional assumptions, such as the absence of isthmuses. Contrary to a plane graph, which has an unbounded outer face, a planar map has no face with a specified state. Planar graphs are those that are capable of being generated on the surface of a particular genus. Planar graphs have genus 0 since the plane and sphere are genus 0 surfaces.

It is maximally planar if a fundamental graph is planar but loses that property when any edge (on the given set of vertices) is added. The term "plane triangulation" refers to the fact that each face, including the outermost, has three boundaries. The terms "triangular graph" and "triangulated graph" are deceptive because they refer to chordal graphs and line graphs of a complete graph, respectively. Every maximal planar graph has at least three edges. A maximum planar network has exactly 3v minus 6 edges and 2v minus 4 faces if and only if there are v vertices with v greater than 2. Triangular faces are divided into triples of smaller triangles to generate Apollonian networks, the most planar graphs possible. It is comparable to planar three-trees. Triangles represent the outermost cycle in strangulated graphs. A maximum planar graph, or more generally, a polyhedral graph, is strangulated because the periphery cycles are the faces. Clique-sums of the most complete and planar graphs can be utilized to generate chordal graphs without removing any edges (Kuchment, 2013).

It is an outerplanar graph attribute that every vertex of an outerplanar graph belongs to the unbounded face of the embedding. K4, however, is planar, but not outerplanar. An outerplanar planar graph is a planar graph. A finite graph is regarded as outerplanar only in the absence of a subdivision of K4 or K2,3. The obvious result of the assertion that an outerplanar graph G can be constructed from G by adding a new vertex and edges connecting it to all existing vertices while retaining the outerplanar property is that G is now an outerplanar graph. A graph's outerplanar embedding and 1-outerplanar embedding are identical. If removing the exterior face's vertices results in a (k – 1)-outerplanar embedding, then the planar embedding is k-outerplanar if k is greater than 1. Graphs are k-outerplanar if their embeddings are also k-outerplanar. A graph with upward oriented and non-intersecting edges is known as an upward planar graph. Because not all planar-directed acyclic networks are upward planar, it is an NP-complete problem to determine whether a given graph is upward planar. It is said that a planar graph is convex if every face, including the outer face, is a convex polygon. Every single planar graph, including the entire bipartite graph K2,4, lacks convex embeddings. A graph cannot be illustrated convexly unless it is a subdivision of a planar three-vertex-connected graph. In addition, Tutte's spring theorem permits the interior vertices of simple 3-vertex-connected planar networks to be positioned as the average of their neighbors (Lengyel, 2011).

Any planar straight-line graph is a planar straight-line graph. The rectangular subset of the integer lattice can be used to construct universal point sets that are quadratic. There exists a universal point set for each vertex of every n-vertex planar network. N-vertex regular polygons provide a universal representation for outerplanar networks because any fundamental outerplanar network permits an embedding in the plane where all vertices are on a fixed circle and all edges are straight line segments that lie inside the disk and do not overlap. According to Scheinerman's conjecture, which is now a theorem, each planar graph can be described as a graph of intersecting line segments in the plane. Each n-vertex planar network can be divided into two subgraphs of size 2n/3 by removing O(n) vertices, according to the planar separator theorem. Treewidths and Branch widths of planar graphs are of order O(n). The strong graph product of an eight-treewidth graph and a route proves that any planar graph is a subgraph according to the planar product structure theorem. This study revealed that planar graphs include restricted queue numbers, bounded non-repetitive chromatic numbers, and linear-sized universal graphs. Additionally, it can be used to color planar graphs and to rank vertex positions.

7.8. HYPERGRAPH

Undirected hypergraphs can be utilized to model satisfiability problems, databases, machine learning, and Steiner tree problems, among other things. They are frequently employed in machine learning tasks as classifier regularization (mathematics) and data model. Examples of applications include image retrieval (correlations as hyperedges), bioinformatics (biochemical interactions as hyperedges), and recommender systems (communities as hyperedges). Two notable hypergraph learning techniques are hypergraph spectral clustering, which uses the hypergraph Laplacian to extend spectral graph theory, and hypergraph semi-supervised learning, which incorporates additional hypergraph structural cost to restrict learning outcomes. For large-scale hypergraphs, an Apache Spark-based distributed framework is also available. Using directed hypergraphs, telephony, operations research, the detection of money laundering, and transportation planning can all be simulated. Additionally, they can be utilized to represent Horn-satisfiability (Kuchment, 2013).

Even though constructing hypergraphs on paper is more challenging than designing graphs, several researchers have investigated techniques for visualizing hypergraphs. In one possible visual representation for hypergraphs, the vertices are depicted as points, disks, or boxes, and the hyperedges are depicted as trees with the vertices as their leaves, similar to the conventional graph drawing style in which arcs in the plane are used to represent graph edges. If the vertices are points, the hyperedges may be represented as smooth curves connecting groups of points or as straightforward closed curves containing groups of points.

The subdivision model of hypergraph drawing is a novel technique for representing hypergraphs that divides the plane into parts, each of which represents a single hypergraph vertex. These areas comprise the hypergraph's hyperedges, which can be colored, outlined, or both. For instance, an order-n

Venn diagram can be viewed as a subdivision drawing of a hypergraph with n hyperedges (the diagram's defining curves) and 2n 1 vertices (the regions into which these curves divide the plane). In contrast to the polynomial-time recognition of planar graphs, determining whether a hypergraph has a planar subdivision drawing is NP-complete; however, the presence of such a drawing can be evaluated efficiently if the adjacency pattern of the regions is constrained to be a path, cycle, or tree. The illustration at the top of this page depicts a variant form of the PAOH hypergraph. Edges are the vertical lines connecting two vertices. The order of vertices is from left to right. On the right, the legend lists the names of the edges. Although it was created for dynamic hypergraphs, it can be applied to basic hypergraphs as well (Lengyel, 2011).

Traditional hypergraph coloring allocates one of a set of colors to each vertex of a hypergraph, ensuring that each hyperedge contains at least two vertices with distinct colors. This implies that a monochromatic hyperedge with at least two cardinalities cannot exist. It is a straightforward generalization of graph coloring in this regard. The chromatic number of a hypergraph is the fewest number of distinct colors used across all colorings. K-colorable hypergraphs can be colored using a maximum of k distinct colors. The hypergraphs with two colors are bipartite. There are several generalizations of the standard hypergraph color scheme. One of them is so-called mixed hypergraph coloring, which permits monochromatic borders. Some blended hypergraphs cannot be colored in any way. There is no universal standard for colorlessness. The lower and upper chromatic numbers specify, respectively, the minimal and maximum number of colors that can be used in a mixed hypergraph (Kuchment, 2013).

7.9. EULERIAN GRAPHS

A finite graph path known as a "Eulerian trail," or "Eulerian route" travels each edge accurately while allowing vertices to be revisited. A Eulerian circuit, also known as a cycle, is a route that begins and ends at the same vertex. Leonhard Euler initially studied them in 1736 while working on the renowned Konigsberg Seven Bridges issue. Euler proved the need for Eulerian circuits, but not that linked graphs with all even-degree vertices must include one. Carl Hierholzer, who died in 1873, was the first to thoroughly reject this latter hypothesis. According to Euler's Theorem, a Euler cycle exists if and only if each vertex in a linked graph has the same degree. In graph theory, the word "Eulerian graph" has two unique meanings. The first is a graph with a Eulerian circuit, whereas the second is a graph with an even degree at each vertex. Connected graphs are defined using the same ideas. Because Eulerian traces are only conceivable when zero or two vertices have an exceptional degree, the Konigsberg graph is not a Eulerian graph. If no vertex has an odd degree, all Eulerian pathways are circuits. Every Eulerian trace begins and finishes at one of two vertices with an odd number of degrees if there are precisely two of them. In semi-Eulerian graphs, there is a Eulerian trace but no Eulerian circuit (Kuchment, 2013) (Figure 7.4).

A Eulerian trail, also known as a Euler walk, is a path in an undirected graph that precisely circumnavigates each edge once. If such a walk is conceivable, a graph is said to be traversable or semi-Eulerian. The Eulerian cycle, also known as a Eulerian circuit or Euler tour, is a cycle that uses each edge precisely once.

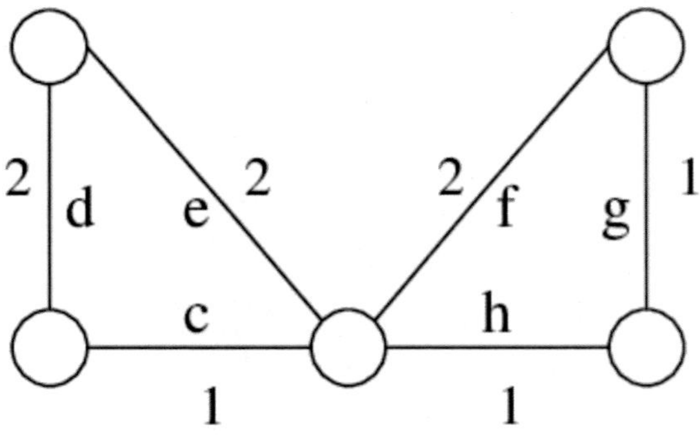

Figure 7.4. A Eulerian graph. The weight and label of each edge is given on the figure.

Source: Teypaz, & Rapine (2008); Scientific Figure on ResearchGate.

If a cycle occurs, the graph is called Eulerian or unicursal. A "Eulerian graph" is a graph in which each vertex has an even degree. A supposedly unconnected graph, on the other hand, is only Eulerian in the weaker meaning if each linked component has a Eulerian cycle. Both formulations are identical for graphs with finite connections. For directed graphs, replace "Path" and "Cycle" with "Directed Path" and "Directed Cycle," respectively. There are several parallels between multigraphs, Eulerian traces, cycles, and graphs. A graph with no discernible orientation and Eulerian orientation the in-degree of v equals the outdegree of v at each vertex, v, and each edge of G is labeled a direction. By generating a Euler tour in each linked component of G and then orienting the edges depending on the tour, this orientation may be constructed for any undirected graph of equal degree. A linked graph that is strongly directed, strongly connected, and extremely Eulerian-oriented (Lengyel, 2011).

According to Euler, for a finite network to be Eulerian, each vertex must have the same degree. Hierholzer determined in research in 1873 that this is a requirement. As a result, the following criterion is both necessary and sufficient for a Eulerian finite graph: When each vertex has the same degree, an undirected, finite, and linked graph is Eulerian. In 1912, Veblen established that a graph is exclusively Eulerian if and only if certain cycles are disconnected (Kuchment, 2013). A non-Eulerian-directed graph with equal degrees is feasible to build. The in-degree and out-degree at each vertex must be equal for a Eulerian circuit to exist. Connectivity is also critical. These conditions are also met, according to Konig. Eulerian graphs are only linked directed graphs with equal in- and out-degrees at each vertex. It makes no difference in this theorem whether "connected" refers to "weakly connected" or "strongly connected," because they are identical in Eulerian networks. Directed graphs may be generated using the linear-time operating Hierholzer algorithm (Lengyel, 2011).

It is not always Eulerian if every degree on a graph is even. This implies that evenness is a required but not sufficient requirement for a mixed graph's Eulerianity. If a graph is both even and symmetric, it must be symmetric. This means that if a graph lacks uniformity and symmetry, it cannot be Eulerian. This criterion is not required since an even, non-symmetric Eulerian graph can be created. In their 1962 book Flows in Networks, Ford, and Fulkerson developed a necessary and sufficient condition for a network to be Eulerian, noting that each vertex must be even and meet the balancing requirement. The number of arcs leaving and entering each subset of vertices S must be fewer than or equal to the number of edges intersected by S. This is a requirement for a balanced set. Only when a graph is balanced and even is it deemed Eulerian. Because the balanced set requirement applies to every possible subset of vertices, calculating the Eulerianity of a mixed graph is more difficult (Kuchment, 2013).

7.10. HAMILTONIAN GRAPHS

According to graph theory, a Hamiltonian path (also known as a traceable path) is a path in an undirected or directed graph that meets each vertex precisely once. Each vertex is traversed precisely once by a Hamiltonian cycle (or Hamiltonian circuit). A Hamiltonian route may be created by deleting any edge from a Hamiltonian cycle. The Hamiltonian cycle and route issues are NP-complete problems that involve finding such cycles and pathways in graphs.

William Rowan Hamilton created the icosian game, commonly known as Hamilton's riddle, which requires the identification of a Hamiltonian cycle in a dodecahedron's edge graph. To overcome this issue, Hamilton created the icosian calculus, an algebraic system based on the roots of unity that has many parallels to quaternions (another Hamilton invention). This approach does not work with random graphs. Despite being named after Hamilton, Thomas Kirkman had previously researched Hamiltonian cycles in polyhedra and presented an example of a polyhedron without them. In the ninth century, the Indian mathematician Rudrata and the Islamic mathematician al-Adli ar-Rumi investigated the Hamiltonian cycles and routes in the knight's tour, commonly known as the chessboard knight's graph. In 18th-century Europe, Abraham de Moivre and Leonhard Euler penned knight's voyages (Lengyel, 2011).

A Hamiltonian route may only be extended into a Hamiltonian cycle if its ends are close. Although a biconnected graph (such as the Petersen graph) doesn't need to be Hamiltonian, all Hamiltonian graphs are. A Euler tour is a closed walk that precisely travels each edge of a Eulerian graph G (a linked graph with the same degree for each vertex). Because this tour corresponds to a Hamiltonian cycle in the line graph L(G), every Eulerian graph has a Hamiltonian line graph. Whether or whether graph G is Eulerian, the line graph L(G) of any Hamiltonian graph G is Hamiltonian. Line graphs may have Hamiltonian cycles that do not follow Euler tours. Hamiltonian tournaments (those with more than two vertices) have a lot of connections. There are (n – 1)!/2 distinct Hamiltonian cycles in a fully undirected network with n vertices, whereas there are (n – 1)! separate Hamiltonian cycles in a fully directed graph with n vertices. These counts, presumably, make no distinction between identical cycles other than their starting position (Kuchment, 2013).

CHAPTER 8

Mathematical Counting and Combinatorics

Contents

8.1. INTRODUCTION

Combinatorics and counting are mathematical disciplines that involve the study of counting and organizing items, combinations, and permutations. They are indispensable in numerous branches of mathematics, computer science, and everyday problem-solving. Counting and combinatorics are utilized extensively in numerous fields, such as probability, statistics, encryption, and graph theory. Possessing this knowledge enables you to solve a variety of problems involving arrangements, options, and probabilities (Quinn, 2022) (Figure 8.1).

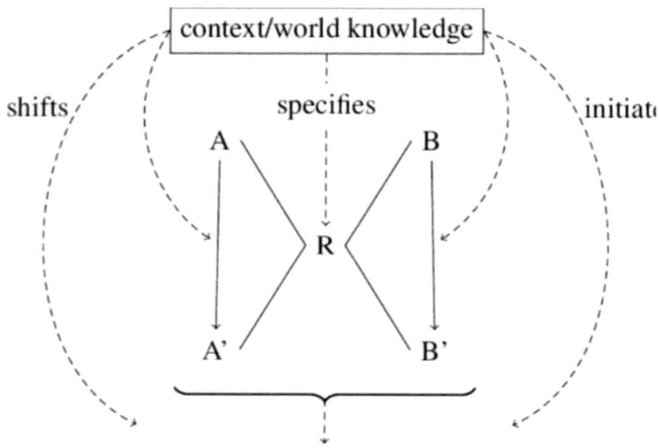

Figure 8.1. Scheme for A B combinatorics.

Source: Bell & Schäfer (2013); Scientific Figure on ResearchGate.

Combinatorics is the scientific study of enumeration and organization. Obviously, the vast majority of humans are capable of counting, but combinatorics employs mathematical techniques to tally objects that are far too large to be counted regularly. Combinatorics has many applications in computer science. It is possible to approximate the number of operations required by a computer program using combinatorial techniques. In the study of discrete probability, combinatorics is also essential. Combinatorics, also referred to as combinatorial mathematics, is a branch of mathematics concerned with selection, organization, and operation in discrete or finite systems. In addition, combinatorial geometry, a closely related subject, is discussed (Mazur, 2022).

Determining the total number of distinct configurations (such as graphs, designs, or arrays) of a given type is a fundamental problem in combinatorics. Despite the relative simplicity of the concepts defining the structure, enumeration can occasionally present significant challenges. The mathematician might have to settle for an approximation or, at the very least, a trustworthy lower and upper bound. In general, an entity is said to "exist" in mathematics if a mathematical example matches the abstract attributes of the entity. In this regard, it may not be immediately apparent that even a single configuration possesses the required properties. This situation is problematic for both existing and new structures. In addition, there is a large class of theorems that guarantee the existence of particular alternatives under valid hypotheses. In addition to their intrinsic value, these theorems can be applied as existence theorems in a variety of combinatorial contexts. Using the potent instrument of combinatorics, mathematicians, and researchers can analyze and quantify the various ways in which items from a given set can be ordered or selected. It provides a systematic method for addressing counting problems and analyzing their underlying structures in a wide range of mathematical and practical applications. In both theoretical and applied mathematics, counting, and combinatorial techniques are utilized to assist in the development of algorithms, data processing, and decision-making (Quinn, 2022).

Counting is a fundamental mathematical operation that entails determining the number of elements or objects in a collection or set. It serves as the basis for numerous mathematical principles and problem-solving strategies. Counting can take many different forms, from simple activities such as counting one's digits to more complex situations involving large sets or combinations.

In mathematics, counting can be performed systematically by cataloguing potential outcomes, utilizing equations, or employing particular combinatorial concepts. In probability, statistics, and a variety of other real-world applications where the number of possibilities or outcomes must be calculated, counting is an essential skill (Mazur, 2022).

Combinatorics is the study of discrete structures and their combinations. It is a mathematical branch. It emphasizes comprehending and analyzing the numerous product categories, options, and divisions. Combinatorics is extensively used to solve problems in a variety of fields and plays a crucial role in comprehending the number of options available in various situations. Combinatorics examines permutations, combinations, and other combinatorial objects like partitions and graphs. Combinatorial techniques are utilized in computer science, cryptography, network design, scheduling, and optimization to solve problems (Quinn, 2022).

8.2. CONCEPTS IN COMBINATORIAL THEORY

8.2.1. Permutations

The term permutation refers to the sequential arrangement of constituents. A group of n unique items has (n) permutations, where n! is the product of all positive integers from 1 to n. Permutations are used when order is essential, such as when arranging individuals in a line or choosing race victors. Combinations are selections of items from a collection that are made without regard to their order. C(n, r) or nCr indicates the number of combinations, which is equal to n/(r!(n-r)!), where n is the number of distinct elements in the set and r is the number of elements selected. Combinations can be utilized to select a committee from a group of individuals or a subset of objects from a collection when the order of the items is irrelevant. Constants of Binary C(n, r) or nCr represent binomial coefficients, which are the coefficients of the elements in the expansion of (a + b)n. They are utilized in algebra, probability, and calculus and have close ties to combinations (Mazur, 2022).

Permutations include all product combinations. n, known as "n factorial," denotes the sum of all positive numbers from 1 to n. It represents the number of permutations for n things. The set contains 1–n positive integers.

"A," "B," and "C" make the ABC sign.

ACB

BAC BCA

CAB CBA

Sequentially rearranging "A," "B," and "C" creates a new order. "P" and the number of pieces or objects rearranged define permutations. The largest number of permutations from "n" items is P(n). Three combinations exist. Factorial notation is used to count "n"-item permutations. n is the mean of all positive integers between 1 and n.

- $4 = 4 \times 3 \times 2 \times 1 = 24$;
- $5= 5 \times 4 \times 3 \times 2 \times 1 = 120$.

The number of permutations is "n!" where "n" is the total number of elements. After an item is picked, there are "n" additional ways to choose it, followed by "n-" ways, "n–1" ways, and so on, until just "1" way remains. Variations persist Many linked classes exist. Imagine "book"

- BOOK;
- BOOK;

- BOKO;
- BKOO;
- BOOK;
- BOOK;
- OOBK;
- OOKB;
- OKBO;
- OKOB;
- OOBK;
- OOKB.

Two "O" spellings may be unnecessary. Repetitions require a permutation formula change. The sixth most common combination always requires both elements. The cardinality of the set of links between r items picked from n objects is P(n, r) or nPr. The factorial function "n" can be used to count the number of "n"-distinct permutations. Permutations can be calculated. For every n/m combination. Thus, if m duplicates and n unique. "N" variants of each item include "m" repeats of the same thing and "p" repetitions of something else (Quinn, 2022).

8.2.1.1. Applications of Permutations

- Cryptography uses permutations to hide data.
- Probability theorists utilize permutations to estimate experiment or event outcomes.
- Combinatorial algorithms use permutations to catalog all possible permutations of its components.
- Natural selection shapes genetic algorithm variations.
- Computer networks use permutations to route data.
- Each video game has its own skill-based rules.

Combinatorics lists every possible combination using permutations. Permutations answer many mathematical and practical issues. Probability, optimization, and statistics need them. Repetition without replacement creates combinations and permutations. Unless the components are ordered, a subset is a combination or a permutation. Pascal and Fermat, two 16th-century French mathematicians, founded combinatorics and probability theory. To assess each subgroup's probabilities of winning random contests, the researchers split the number of subgroups that interested them (Mazur,

2022). Five pieces can be joined from A to E. All 20 alternatives may be verified using the letter sequence and proposed order. Permutations are set-element combinations. This enumeration lists all set components. Two groups result from A = 1. Set A has stationary items. Combinations are less sequence-sensitive than permutations. Permutations have three kinds. The first cycle randomizes N items. Permutations can exchange items if there is no unique one in mathematics. First, list all permutations. P(n, r) gives a rough estimate of the number of permutations when picking r times from a set of n distinct items. Time reduces permutations to zero. P(r) is n–1 independent of r.P(n, r) is the probability of an R-value association. (n – 3)……(n – r + 1) "nPr" displays the combinatorial computation of the number of permutations that may be generated from a collection of "n" objects when "r" are selected at random and the order of selection matters (Quinn, 2022). Permutation computations are automated after a certain number of iterations. Exponents represent predicted variability. If each object has r versions, you can choose n of them. Repetition might vary. (r x) = nr. Taking into account recurrence, the permutation formula predicts the number of combinations that may be constructed by picking r components from n items. If repetition is allowed, the maximum number of three-letter words from "smoke" must be established.

8.2.2. Combinations

On the other hand, pairings are simply picked at random. The number of possible permutations, denoted by C(n, r), is equal to n!/(r!(n-r)!). The properties of combinations and permutations are quite similar. However, the order of items is necessary for permutations. Combinations of ab and ba are the same, while permutations of ab and ba are not. Combinations have many applications in mathematics, finance, and combinatorics. The mathematical procedure for counting the variety of possible combinations from a small sample. Follow these steps if you want to choose two of the three choices (A, B, or C):

AB

AC

BC

Since the order of elements does not matter in combination, AB = BA. Mathematicians use the concept of combinations to count the number of random ways that a certain number of items may be chosen from a bigger set. The emphasis in combinations is on "selecting" rather than "arranging" components. Choose r at random from the available n options. Produce like apples, avocados, oranges, and pears is typical. In what ways may fruits and vegetables be combined? C(n, r) or nCr represents a combination where "n" is the total number of unique elements in the set and "r" is the desired size of the set (Quinn, 2022). Counting combinations using the equation r (n-r)/n C(n, r) = r. The sum of all positive numbers between 1 and n is (n-r)/n n. A factorial of r is equal to the sum of all positive integers from 1 to r. (n – r) the (n – r) factorial consists of all positive numbers from 1 to (n-r). Let's go through all the many ways we may select two fruits from the set "Apple, Banana, Orange, and Pear" again (Mazur, 2022).

It's a four-veggie n. Two servings of veggies, please.

C (4, 2) = 4/(2!(4 – 2)) = 24/(2 × 2) = 6 is the final answer.

Choose two fruits from the selection below: pears, apples, bananas, or oranges.

Bananas with apples

Fruit juice and apple

Citrus, Banana, and Pear

No matter what sequence you put the components in, the result will be the same. Apples and bananas are interchangeable. As a result, their identities are combined. Combinations are used in fields like probability and statistics where the order of choices doesn't matter, such as when addressing a problem.

8.2.3. Permutations with Repetition

There is a special formula to get the total number of permutations that include duplicates of the same items. n/(r1 × r2 ×… × rk). It's easy to find the permutation of repetitions. The number of permutations for an item having n kinds and r choices is n n… (r times). The number of choices doubles with each succeeding option, thus there are n choices for the first option, n choices for the second, and so on.

The arrangement of similar or interchangeable things is achieved through repetition and permutation. Therefore, a special approach is required for tallying the permutations of components that repeat themselves. Take this as an illustration. Contemplate the word "BOOK."

- BOOK;
- BOOK;

- BOKO;
- BKOO;
- BOOK;
- BOOK;
- OOBK;
- OOKB;
- OKBO;
- OKOB;
- OOBK;
- OOKB.

There are some repetitions in the permutations because the word has two 'O's. To take into account repeats, the permutation formula has to be adjusted. For a set of n items that are repeated r1, r2, …, rk times, we may get the total number of permutations (P) as follows: (r1 × r2 ×…× rk) The sum of all positive integers between 1 and n is P = n n, sometimes called the n factorial. The repeated factorials are [r1, r2,…, rk] (Quinn, 2022).

There are P possible permutations in our "BOOK" example.

$P = 4/(2 \times 2) = 24/(2 \times 2) = 6$

There are six different permutations of the word "BOOK" due to the double occurrence of the letter "O." When the constituent elements are all the same, it becomes necessary to repeatedly permute them. They have applications in the fields of algorithm design, cryptography, combinatorics, and probability (Mazur, 2022). Vandermonde's identity connects binomial coefficients and combinatorial repetitions. A multinomial's components are the same as the sum of its coefficients. Repeated permutations are emphasized in sentence 8. Putting things in a straight line, like books on a shelf, might help you notice patterns. There are more conceivable arrangements with more identical objects. The following are examples of permutations in P(4; 1, 2, 1): repeated sequences of "BBAYB," "BYABB," etc., where B1, B2, and B3 are the three "B's," A is the single "A," and Y is the single "Y."

8.2.3.1. Features of Repeated Permutations

The total number of permutations grows as more and more copies of the same elements are used. Combinations are made up of random picks, as opposed to the methodical ordering of permutations. In combinatorics, repeated permutations focus on elements that appear repeatedly. Multiple combinations may be easily calculated using the provided formula and attributes. This idea recycles the same stuff in different contexts to solve real issues.

8.2.4. Combinations with Repetition

Repetition-Based selecting numerous items from a group requires the use of combinations. Repeated combinations have the form $(n + r - 1)$. Where n is the number of items and r is the number of possible choices for each object. To count the possible permutations while selecting from a set where duplicates are permitted, "stars and bars" or "balls and urns" puzzles are used in the field of combinatorics. This issue occurs when we have a set of distinct items and want to count the number of possible permutations that may be made by choosing some of them, keeping in mind that the same thing can be chosen more than once (Quinn, 2022).

To restate, we'd want to select r among n unique goods. Think about the alphabet and the fruits apple (A), banana (B), and orange (O). Pick five types of veggies. If you can choose the same fruit more than once, how many different ways are there to combine them?

This issue is resolved by the formula for recursive combinations $(C(n + r - 1, r))$.

In this situation, n represents the total number of distinct citrus types.

The total number of goods to collect is r. A, B, and O are the names of our three citrus types. (Select 5 Veggies.) $r = 5$.

$C(3+5-1, 5) = C(7, 5)$.

Determine $C(7, 5)$. Number of permutations $(nCr) = r(n-r)/n$ The formula for $C(n, r)$ is $r(n-r)n$

These are the solutions to this equation: $C(7, 5) = 7/(5(7 - 5)! = 7!/(5 \times 2) = (7 \times 6 \times 5 \times 4 \times 3 \times 2 \times 1)/((5 \times 4 \times 3 \times 2 \times 1) \times (2 \times 1)) = (7 \times 6)/(2 \times 1) = 42/2 = 21$

If you pick five different fruits from the letters A, B, and O again, you'll end up with 21 different permutations. Comparable combinations to:

AAAAA

ABBBA AABBA BBBBB BBBAA AOAAA

Recognizing Iterative Patterns: $C(n + r - 1)$

$C(n + r - 1)$ There are r iterations in total.

- n = the total number of unique things.

You have r chances to choose each option. Combinations Based on Repetition "Stars and bars" allow for the depiction of such combinations. Think about "n" objects and "r" spots. We call the coordinates "bars," and each individual item is a "star." A, B, and O are the fruits you must choose five of. There were three "bars" in between the five "stars":

OO|OOO|A|A|B

No bananas, but there are two oranges (OO), three apples (OOO), an A, and a B. The selected crops are represented by the stars, while the "divisions" are created by the bars. This pattern of "stars and bars" is used to represent recurrence in a special way (Quinn, 2022).

5. Formula Application: Toss in five more of each: apples, avocados, and oranges. The equation deduces the following iteration combinations:

$C(3 + 5 - 1, 5) = C(7, 5) = 7!/(5! \times 2!) = (7 \times 6 \times 5 \times 4 \times 3 \times 2 \times 1)/((5 \times 4 \times 3 \times 2 \times 1) \times (2 \times 1)) = 21$

If you pick five different fruits from the letters A, B, and O again, you'll end up with 21 different permutations.

8.2.4.1. Applications of Combinations with Repetition

- When the same combinations are distributed again, they are placed in many different buckets. Treating kids to candy, assigning everyone the same work, and funding initiatives are all good examples (Mazur, 2022).
- Tally duplicate items in subsets by combining them again.
- Probability and statistics: Recurring combinations are used to evaluate the probability of specific outcomes in repeated circumstances.

"Balls and Urns" is a common idiom. These puzzles usually include many permutations and repetitions, including the placement of identical spheres in a set of urns. To produce every feasible combination of objects with repetitions, combinatorial algorithms are needed. In order to count the number of unique combinations that may be made from a set of duplicates, combinatorics makes use of repeated configurations. The phrase "stars and bars" shortens the discussion. Problems in probability, statistics, distribution theory, and combinatorial algorithms may all be tackled with the use of combinations with repetition. Real-world examples of repetition include the organization of like items into bins, the construction of uniformly sized teams, and the maintenance of inventory lists. Methodical answers are needed for these issues in the fields of probability theory, statistics, computer science, and optimization. Numerous difficulties may be solved with the use of knowledge of combinations and combinations with repetition (Quinn, 2022).

8.2.5. Binomial Coefficients

C(n, r) or nCr denotes the binomial coefficient, which is a mathematical representation of the number of ways r items may be chosen from a collection of n components. Permutations and combinations have a strong link and are widely used in probability theory, algebra, and calculus. Binomial coefficients, or binomial theorem coefficients, are always positive integers. The words "binomial coefficient" and "combinatorial number" are two additional names for combinations. The sum of all potential outcomes contained inside a particular collection may be calculated using combination and permutation approaches. Combinations and permutations handle order in fundamentally different ways. The sequence of events is unimportant in combinations, but it is critical in permutations. The goal of this research is to look at a wide range of binomial coefficients (Mazur, 2022).

Alternatives to "n rather than k" that may be beneficial include "n prefer k" and "n instead of k." One of its connected meanings is represented by the symbol $binomnk$. As a result, these events are sometimes described using combinations and combinatorial numbers. The following discussion will look at the properties of binomial coefficients. To obtain these numbers, we employ Pascal's triangle, which was devised by mathematician Blaise Pascal. A combinational analysis is being tried, and assistance is required. According to the binomial theorem, the nth power of a positive number n may be represented as the sum of terms of the form n+1.

A unique pattern appears when the exponents of the coefficients in a binomial equation are taken to their maximum values. There are several mental frameworks that correspond to these sentences (Quinn, 2022).

There is one more word in each power of two greater than the binomial. Multiplying the exponents of the terms in a binomial produces the same result as multiplying the exponent of the binomial itself.

Skills designated by the letter 'a' in the expansion's word list lose one competency level, whereas skills denoted by the letter 'b' gain one level. A symmetrical framework is provided by the coefficients. The coefficient item in the second row reflects the total of each pair of integers. Pascal, a

French mathematician, named the legendary triangle. Binomial coefficients, often known as fundamental combinatorial numbers, have a wide range of applications in mathematics. Methods for calculating term coefficients in binomial expansions are presented. When both "n" and "r" are positive integers, coefficients can be represented as C(n, r) or nCr.

The first step is to conduct research on the subject matter. The technique of doing so is known as extending a binomial equation. Binomial expansions are algebraic expressions that involve binomials increased to a positive integer exponent. In a binomial equation, the plus or minus sign serves as a separator. A binomial equation is represented by the phrase (a + b), where the terms "a" and "b" can be either variables or constants.

Where "n" is a positive integer, the binomial expansion of the equation (a + b)n is (a + b)n = C(n, 0), where C(n, 0) is the one-at-a-time combination of n objects… + C(n, n) As seen below, the formula may be stated as a multiplied series of terms, with each term being the product of a variable "a" raised to a power and a variable "b" raised to a power: a × b0 × C(n, 1) × a × b2 × C(n, 2). The numbers for "a" can range from 2 to 0, whereas the values for "b" can range from 0 to "n" in power.

The binomial coefficient C(n, r) is used to denote the word at index "r" in the expansion. The sequence number represents the combinatorial count of selecting a subset of r components from a collection of n items. The binomial coefficient is calculated as follows: n!/(r!(n-r)!. The formula for computing the number of combinations is C(n, r), which has the notation n!/(r!(n-r)!). The binomial coefficient is defined as the product of all positive integers from 1 to "n" divided by the product of all positive integers from 1 to "r" multiplied by the product of all positive integers from 1 to "n-r." The binomial coefficient has various advantageous properties.

The exclusion of "n – r" components from a set of "n" is identical to the exclusion of "r" members from the same set for symmetric binomials, resulting in the equivalence C(n, r) = C(n, n – r).

Pascal's Triangle is a triangular array of integers initially proposed by the French mathematician Blaise Pascal in 1787. Pascal's Triangle is formed by arranging the coefficients of binomial expansions in a triangle arrangement, with the first and final elements of each row assigned the value 1, and the remaining elements decided by the sum of the preceding row's two adjacent values (Quinn, 2022).

The equation C(m + n, r) = (C(m), k) × C(n), r – k) is valid for all k values between 0 and r.

The Binomial Theorem is a source of contention. The binomial theorem may be used to conduct a variety of mathematical operations, such as calculating the digits or computing the remainder of a number, which is why it is so often employed. The remainder may be calculated using the Binomial Theorem (Mazur, 2022).

We are looking for the quotient produced by dividing 7^{103} is divided by 25. Using these data and the remainder formula, we can calculate:

$(7^{103/}25) = [7(49)^{51}/25)] = [7(50 - 1)^{51/}25]$

$= [7(25K - 1)/25] = [(175K - 25 + 25 - 7)/25]$

$= [(25(7K - 1) + 18)/25]$

Determine what value of K must be utilized in order for $(2^{403}/15)$ is $(K/15)$, then find K.

Here are some possible solutions to this problem:

$(2^{403}/15) = [2^3 \, (2^4)^{100}/15]$

$= 8/15 \, (15 + 1)^{100} = 8/15 \, (15\lambda + 1) = 8\lambda + 8/15$

$\because 8\lambda$ is an integer, fractional part $= 8/15$

So, $K = 8$.

$(13)^{10} = (169)^5 = (170 - 1)^5$

$= 5C_0 \, (170)^5 - 5C_1 \, (170)^4 + 5C_2 \, (170)^3 - 5C_3 \, (170)^2 + 5C_4 \, (170) - 5C_5$

$= 5C_0 \, (170)^5 - 5C_1 \, (170)^4 + 5C_2 \, (170)^3 - 5C_3 \, (170)^2 + 5(170) - 1$

A multiple of $100 + 5(170) - 1 = 100K + 849$

$\therefore$ The last two digits are 49.

The cardinality of the set of variables in the preceding expansion was equivalent to the cardinality of the set of non-negative integral solutions (Mazur, 2022).

Because each solution adds a term to the overall expansion, we may select the one that best matches the data. Coefficients based on the binomial distribution are widely employed in a variety of mathematical applications.

In the study of combinatorics and counting issues, layout, and component selection are critical; binomial coefficients are a mathematical representation of the different combinations from which r items can be picked from a given collection of n items (Quinn, 2022).

The binomial distribution is a mathematical model that quantifies the likelihood of achieving a certain number of successes in a fixed number of independent Bernoulli trials using the concept of binomial coefficients, which are used in probability theory to determine the likelihood of an event occurring within a specific statistical domain known as the sample region.

Binomial coefficients are commonly used in mathematics for a range of simple algebraic operations such as expansions and simplifications. Polynomials are important in calculus because they allow for the production of approximation functions, which are required for the application of Taylor series and power series in function expansion. Binomial coefficients are commonly used in algorithms for evaluation and optimization, and their use is critical in the field of computer science due to the frequency of challenges involving combinations, permutations, and probability. Binomial coefficients are commonly used in a variety of mathematical and practical applications due to their utility in illustrating the binomial expansion. This curriculum provides a thorough introduction to problem solving in algebraic equations, probability theory, and option pricing.

8.2.6. The Pigeonhole Principle

A basic counting rule known as the "Pigeonhole Principle" asserts that if there are n pigeonholes and there are more than n things to place into them, then at least one pigeonhole must contain more than one object. Numerous disciplines, including computer science and number theory, use this idea. According to the pigeonhole principle, at least one container must hold more than one item if n things are placed into m containers, where n > m.[1].

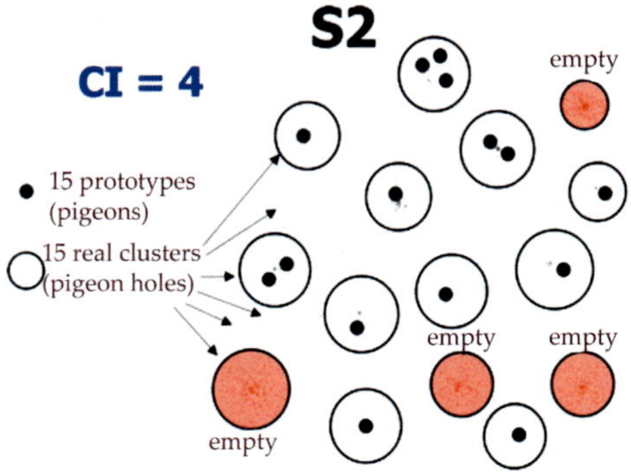

Figure 8.2. Illustration of the clustering problem using a pigeon-hole principle with dataset S2.

Source: 2018; Scientific Figure on ResearchGate.

As an illustration, if one owns three gloves (none of which are ambidextrous or reversible), at least two of them must be right-handed or left-handed as there are three items but only two categories of handedness to classify them into. It is possible to establish potentially surprising consequences using this seemingly straightforward assertion, a form of counting argument. For instance, the pigeonhole principle states that there must be at least two Londoners with the same number of hairs on their heads given that the population of London is more than the maximum number of hairs that may be present on a human's head (Mazur, 2022) (Figure 8.2).

Although the pigeonhole principle first appeared in a text attributed to Jean Leurechon in 1624, it is most usually referred to as the Dirichlet's box principle or Dirichlet's drawer principle since Peter Gustav Lejeune treated the concept in 1834. Schubfachprinzip, sometimes known as the "drawer principle" or "shelf principle," was proposed by Dirichlet. A fundamental idea in mathematics known as the "Pigeonhole Principle" offers a straightforward yet effective method for deducing allocations and counting. It says that if you wish to put more "pigeons" (or things) into "pigeonholes" (or containers), at least one of the pigeonholes must hold more than one pigeon. To put it another way, if you try to distribute objects such that each has a distinct container of their own, but there are more objects than containers, then certain containers will unavoidably contain more than one item (Quinn, 2022).

The concept is intuitively simple to understand and has several real-world applications in mathematics, computer science, probability, and cryptography. Let's delve more into the Pigeonhole Principle:

Imagine having nine pigeonholes (H1, H2, …, H9) and 10 pigeons (P1, P2, …, P10). Because there are more pigeons than pigeonholes, if you try to put each one into a separate one, one of the pigeonholes will be left vacant. However, you must put more than one pigeon in at least one pigeonhole if your goal is to distribute the pigeons so that every pigeonhole has at least one bird. Let's construct a few phrases to formalize this:

- N: The quantity of items, or pigeons;
- M: The quantity of containers, or pigeonholes.

This is how the Pigeonhole Principle is put into words:

If $N > M$, then there must be more than one pigeon in at least one pigeonhole. Regardless of the characteristics of the items or the containers, this concept is valid. It is simply based on a comparison between the quantity of things and containers (Mazur, 2022).

1. **Comprehending the Principle:** The Pigeonhole Principle is based on the straightforward notion of counting items and containers. If there are "m" pigeonholes and "n" pigeons (objects), and "n > m," then there must be at least one pigeonhole that holds more than one pigeon. This is so that we won't run out of holes before we can fit all the pigeons if we try to place each one into its own hole. Therefore, more than one pigeon must be present in certain holes.
2. **Making the Principle Visible:** Consider selecting a pair of socks from a drawer that contains 10 pairs. To ensure that you have a pair of socks that are the same color if the socks are all different colors, you must choose at least two socks. In this illustration, the socks stand in for the pigeons, and the various colors are the pigeonholes. You must choose at least two socks in order to locate a pair, which indicates that you have put two pigeons in one pigeonhole.
3. **Principle General Statement:** The Pigeonhole Principle can be expressed in formal terms as follows:

If there are "n" things and "m" containers, and "n > m," then there must be more than one item in at least one container. This assertion is potent since it is applicable regardless of the characteristics of the items or containers. It is simply based on a comparison between the quantity of things and containers (Quinn, 2022).

8.2.6.1. Applications of the Pigeonhole Principle

1. **Birthday Contradiction:** There is a better than 50% probability that two persons will share the same birthday in a group of just 23 people. The Pigeonhole Principle provides the foundation for this unexpected outcome. The theory predicts a high probability of at least two persons having the same birthdate with 365 potential birthdays (pigeonholes) and 23 people (pigeons).
2. **Cryptography:** The Pigeonhole Principle is used in cryptography to comprehend the likelihood of collisions in hash algorithms. The theory states that collisions (two inputs yielding the same output) are unavoidable for certain functions as hash functions transfer a huge input space to a smaller output space (like stuffing pigeons into pigeonholes) (Mazur, 2022).
3. **Voting and Elections:** Voting processes or elections are subject to the guiding concept. There will always be at least one applicant

who is unsuccessful if there are more contenders (pigeons) than open slots (pigeonholes).

4. **Mathematicians and Researchers:** These can use the Pigeonhole Principle to solve problems and draw inferences about the distribution of things and containers. One may approach hashing algorithms, probability puzzles, and counting difficulties with a clearer and more methodical attitude if they are aware of this notion. The idea may be used to demonstrate how any lossless compression technique, as long as it reduces certain inputs (as the name compression implies), will also increase some inputs. The pigeonhole principle rules out the idea that the set of all input sequences up to a particular length L might be mapped to the (much) smaller set of all sequences of length less than L without collisions (since the compression is lossless).

8.2.7. Factorials

Factorial analysis has applications in both combinatorics and counting. Because it is the sum of all positive numbers from 1 to n, the factorial of n is represented by "n!" Probability theory, permutations, and combinations are only a few of the many mathematical concepts that use factorials. This in-depth investigation will increase your understanding of factorials and their possible applications in addressing mathematical difficulties.

If n is a positive integer, the factorial is the sum of all positive integers from 1 to n. Because the factorial of 5 is 120, the result is 120 if n equals 5. The traditional value for the factorial of 0 is 1. Factorials are the foundation for permutations, which involve the arrangement of unusual objects. Consider a collection of n distinct items. These elements can be mixed in an infinite number of ways, where n is a positive integer. According to the factorial formula 3! = 321, there are six distinct combinations of the three supplied items (A, B, and C).

ABC ACB were the letters inputted by the user.

They abbreviated their surnames to "BAC" and "BCA."

It is controversial which letter comes first, CAB or CBA.

Permutations can be used to arrange items in a certain order. Permutations have several applications, including seating layouts, shelf organization, and data storage. Combinations are built using factorial calculations, which choose a subset of things without regard for their order. The mathematical term "n choose r," or nCr for short, represents the number of permutations. To get this number, multiply the product of the factorials of r and (n-r) by the factorial of n (Quinn, 2022).

When selecting two elements from a set of A, B, and C, the combination formula may be used to compute the number of alternative permutations. Due to the factorial nature of the equation, 3C2 = 3!/(2!(3–2) reduces to three. Among the countless possibilities are:

Given inputs are AB, AC, and BC (Mazur, 2022).

A committee or team may be picked from a larger pool of applicants based on a variety of factors. Probability theory is the study of the mathematical concepts and ideas that underpin the chance that events will occur. Factorials are used in probability theory to tally the possible outcomes of an experiment. Consider a study that has n separate components. These results can be reordered and sequenced n times.

Let us now calculate how frequently a certain event will occur, especially how frequently a 3 will be rolled on a six-sided die. By dividing the number of positive occurrences (getting a 3) by the entire number of probable outcomes (rolling a dice), we arrive at a probability of 1 in 6. This claim is valid since a typical six-sided die has just one outcome. 1, 2, 3, 4, 5, and 6 are the most likely outcomes. All possible combinations of the digits 1 through 6 are presented. By counting both the desired and likely outcomes, factorions make it simpler to evaluate probabilities in circumstances with more complicated probability.

Researchers developed combinatorial algorithms to handle problems with several possible outcomes. The algorithms are put to good use. Factorials are an important concept in combinatorial algorithms, which employ them to generate and count combinations and permutations. Algorithms are utilized in a variety of fields such as computer science, cryptography, and optimization. A recursive factorial method may generate every imaginable combination of the input components (Mazur, 2022).

Counting principles (or counting rules) are basic concepts in combinatorics that give a methodical technique to count the number of objects in a collection. Mathematical factorials rely on the multiplication and division principles. The Multiplication Principle states that the total number of possibilities for a series of unrelated occurrences may be estimated by multiplying the probability at each stage. To identify consecutive possibilities, factorial calculations are typically necessary. A counting issue with many phases and comparable arrangements may be split to find the total number using the Division Principle. The concept of dividing factorials may be demonstrated here. In mathematics, recursive definitions can be used to define a concept or an object. Using recursion, the factorial function may be defined as follows: The factorial of (n–1), denoted by the sign n, is generated by multiplying n by itself for all positive integers n. Using a recursive method makes it easier to understand factorials and solve related problems (Quinn, 2022).

8.2.7.1. Real-World Factorial Applications

Outside of pure mathematics, factorials have various practical applications:

- Factorization is used in sorting, DP, and constructing permutations and combinations algorithms.
- In cryptography, factorials are used to assess the complexity of algorithms and the number of possible keys.
- Permutation testing, and experimental design are all statistical procedures that heavily rely on factorials.
- When working with elements, particles, and energy levels, physicists, and engineers regularly employ factorials.
- Factorials can be used to calculate probabilities and expected values in the realm of uncertain economics and finance.

The concept of factorials is essential in combinatorics and other disciplines of mathematics. Because it is the sum of all positive numbers from 1 to n, the factorial of n is represented by "n!" Probability theory, permutations, and combinatorial algorithms are all uses of factorials. Information technology, cryptography, statistics, and engineering all make use of these specialized subjects. Mathematicians, scientists, engineers, and researchers employ factorials to solve mathematical and real-world problems (Mazur, 2022).

8.3. COMBINATORICS APPLICATIONS

Combinatorics is used in most fields, including:

- Network security, cryptography, and communication networks;
- The study of molecules with computers;
- Scientific research on computer architecture;
- Languages;
- Pattern recognition;

- Simulation;
- Data mining and databases;
- Homeland protection.

Combined operations research solved illustrations.

Arrangement of the letters in the word TAMIL in Example 1

T always follows L.

T and L are inseparable diagnosis.

Let's treat LT as a single letter and retain it intact. Now that we have four alphabets, we may arrange them in a row (Quinn, 2022).

${}^{4}P_{4} = 24$. (Basic formula: nPr)

There are two ways to switch the locations of L and T. As a result, the total configurations are 4!2! = 48. Calculate the variety of ways a cricket 11 may be chosen from a group of 15 players in Example 2 if;

- No limitations on what can be chosen. There is always one player chosen (Mazur, 2022).
- Never is a certain player picked.

Solution: When there are no limitations on the choice. Thus, $15c_{11}$ provides you with the complete number of ways. Since the same player is always chosen. Consequently, 15 – 1 gives us $14c_{10}$. No particular player is ever chosen; therefore, we have $14c_{11}$. Determine how many committees of five students may be selected from a class of 25.

Solution: Since we must choose 5 from a possible 25. Therefore,

${}^{25}C^{5}$=25X24X23X22X21/5X4X3X2X1

= 53130

Combinatorics has several uses in the field of probability. For instance, we may use the following formula to determine the likelihood that event A will occur: P(A) is the probability that an event happens and equals the number of alternative outcomes in which it occurs. Combinatorics may also be used to determine the total number of outcomes.

Example 4: While playing a game, P, Q, R, and S sit at random at the four corners of the classroom. Calculate the likelihood that Q is seated in the north-east corner of the space. We are aware of 24 possible arrangements for students to sit in the corners of a classroom. There are also six (that is, three!) different ways in which Q might occupy the northeast corner. Thus, the likelihood that Q will be seated at the north-east corner of the room is equal to the number of possible ways for Q to do so divided by the total number of possible ways, which comes to 6/24 or ¼ (Mazur, 2022).

In conclusion, counting and combinatorics are branches of mathematics that study the structure of systems. Discrete math, probability, and statistics can't exist without these areas of study. Combinatorics, as contrast to counting, investigates alternative methods to combine outcomes or organize data. According to the multiplication principle in counting, if there are 'n' ways to execute one activity and 'm' ways to complete another, then there are 'n x m' ways to accomplish both. You may choose from 12 unique dishes, including 4 appetizers and 3 sweet finishes.

Inclusion-exclusion can simplify complex counting problems. Mathematicians can determine how many sets have a common union by contrasting their sizes and crossings. Grouping people, enclosing things, and forming candidate committees are all problems that may be solved by the study of combinatorics (Quinn, 2022).

CHAPTER 9

Discrete Probability in Mathematics

Contents

9.1. INTRODUCTION

Probability theory is a key part of math that helps us understand uncertainty and make smart choices. Probability theory helps us understand and calculate the chances of things happening, like guessing whether a coin will land heads or tails. It also helps us predict the chances of things like diseases spreading. In the wide field of probability theory, discrete probability is important for studying and analyzing random events that have a specific number of possible outcomes (Shiryaev, 2016) (Figure 9.1).

Sum	Posible Casess	Favourable Cases	Probability
2	(1,1)	1	1/36
3	(1,2), (2,1)	2	2/36
4	(1,3), (2,2), (3,1)	3	3/36
5	(1,4), (2,3), (3,2), (4,1)	4	4/36
6	(1,5), (2,4), (3,3), (4,2), (5,1)	5	5/36
7	(1,6), (2,5), (3,4), (4,3), (5,2), (6,1)	6	6/36
8	(2,6), (3,5) (4,4), (5,3), (6,2)	5	5/36
9	(3,6), (4,5), (5,4), (6,3)	4	4/36
10	(4,6), (5,5), (6,4)	3	3/36
11	(5,6), (6,5)	2	2/36
12	(6,6)	1	1/36
Total		36	

Figure 9.1. Probability distribution for the sum of two dice.

Source: Burgos, Batanero, & Godino (2022); Scientific Figure on ResearchGate.

Discrete probability is about studying random experiments where the results can be represented by a limited or countable number of specific values. This helps us understand how things behave when they happen randomly and figure out the chances of different outcomes in many different situations. Knowing about discrete probability is important for anyone who wants to understand things like statistics, computer science, making decisions, and measuring risk.

9.1.1. The Importance of Probability

Probability theory is very important in our lives because it helps us understand and deal with the uncertainty that is always there. Probability theory helps us figure out how likely different things are to happen. We can use it to make decisions, look at data, and predict what might happen. By assigning chances to different results, we can make wise decisions, evaluate dangers, and have a better understanding of the world we live in. Let's understand why probability is important in different situations (Epp, 2010).

- Decision-making is when someone has to make a choice or decide something. When you don' have enough information to make a decision, it's important to consider the likelihood of something happening. Sometimes in our personal or business lives, we come across situations where we are unsure about the outcome and have to make decisions. Using probability theory helps people determine the likelihood of different results and choose the actions that align with their desired outcomes. For example, in finance, investors use probability models to figure out how risky different investment options are and how much money they might make. This helps people choose the best investments for their portfolio (Shiryaev, 2016).
- Risk assessment is the process of examining how probable it is for something to go wrong and how serious the consequences could be. Insurance is a deal that provides financial protection if specific events occur. Probability theory is the subject that looks at how likely events are to happen. It is all about understanding risks and handling insurance. Insurance companies use math to figure out how likely something bad happening is and how much money they should ask people to pay for protection. Insurance companies analyze past events and use statistics to predict the likelihood of specific events happening, like car accidents or natural disasters. After that, they choose how much insurance they will offer and how much they will ask people to pay for it. Probability theory helps people and businesses understand and control risks. This helps them make smart choices about keeping themselves safe, investing their money wisely, and using their resources efficiently (Epp, 2010).
- Statistical analysis is a way to study data and make important discoveries. Data science is a field where you use numbers to study problems and find answers. Probability theory is used in statistics and data science to study and understand data. Probability distributions and statistical models are tools that help researchers and analysts study data, make educated guesses, and reach important findings. Statisticians use probability theory to make informed predictions about certain things in a group. After coming up with their ideas, they try them out and see how sure they feel about the outcome. Probability theory is a method that computers use to learn patterns and make predictions. It uses

models that rely on the chances of particular results happening (Shiryaev, 2016).

- Making educated guess and predictions about the future using data analysis and statistical techniques. Probability theory helps us make guesses and predictions in different areas. Weather forecasting uses models to make predictions about future weather conditions. These models take into account past weather patterns, current data, and scientific knowledge about whether to make the predictions. Likewise, in economics and finance, probabilistic models help to predict how the market will change, the prices of assets, and important economic measures. By knowing the possibilities of different results, decision-makers, and companies can make better predictions and prepare accordingly.
- Scientific research is a study that scientists do to gather information and understand different aspects of the world. Probability theory is an important part of scientific research. It helps scientists measure uncertainty, plan experiments, and make trustworthy conclusions. In subjects like physics, biology, and environmental sciences, the use of probabilistic models can help us understand complicated events and make predictions. Probability theory helps us to understand and analyze the random and varying behaviors we observe in natural systems, like quantum mechanics and population dynamics (Epp, 2010).
- Everyday Life are the normal things we do every day. Probability theory is used in many areas that are not specific or limited to certain subjects. It is also applicable in our daily lives. From choosing the best way to get to work, to picking a restaurant using online reviews, we naturally think about the chances of things happening and things that might not be certain. Probability theory is used to understand the chances in gambling and games of chance. It also helps us make smart choices about our health by considering risk factors, and it helps us understand social things by studying survey data and opinion polls.

In simple words, probability theory is very important because it helps us make informed decisions, understand risks, and get information about the uncertain world we live in. It is used in many different areas, like finance, science, statistics, and everyday life. By using probability theory, we can deal with not knowing something more easily, improve how much we know about the world, and make better decisions based on good information (Epp, 2010).

9.2. BASIC CONCEPTS IN DISCRETE PROBABILITY

In math, it's really important to understand and measure uncertainty. Probability theory helps us understand and deal with the uncertain nature of the world. Discrete probability is an important part of probability theory. It is about situations where we get different results, and each result is unique and separate. By learning the fundamental ideas in discrete probability, we can build a strong base to understand and study the patterns and outcomes of these random events (Shiryaev, 2016).

9.2.1. Sample Space and Events

In probability theory, we use the ideas of sample space and events to help us understand and measure uncertainty. They help in organizing and sorting the different potential results of a random test. Sample Space is a term used in probability theory. It refers to the set of all possible outcomes in an experiment or event. In simpler terms, it is the range of all the different things that could happen. The sample space is a collection of all the different things that could happen in a random experiment. This means all the possible outcomes that can happen. The different results that could happen are usually shown by using different values or symbols that cover all the possible outcomes. For instance, when you toss a fair coin, the possible outcomes are only two: either it lands on Heads or Tails. The sample space is very important because it helps us understand what we are studying and lets us organize and classify different outcomes or events. It helps us think about all the possibilities and is used to calculate probability (Epp, 2010).

Events are different groups or sets of possible results that make up the sample space. They let us concentrate on specific events or situations within a bigger group of possibilities. Events can vary from simple things that happen, like when a specific number shows up, to more complicated situations with many different things happening together. Events are represented by big letters like A, B, or C, and are categorized based on how they relate to the sample space. There are various kinds of events.

1. **Simple Events:** These events include only one possible outcome from the collection of all possible outcomes. Let's say you roll a normal six-sided die, which has numbers 1, 2, 3, 4, 5, and 6 on its

sides. If we focus on one specific event, such as rolling a 3, that is called an elementary event.

2. **Compound Events:** These are made up of more than one possible result. They are created by combining basic results together using actions like adding, taking away, or comparing. For example, if you roll a die and get a number 2, 4, or 6, that would be called event A.
3. **Certain Event and Impossible Event:** In simple terms, the certain event means that no matter what happens, something will definitely occur. On the other hand, the impossible event means that there is no chance of something happening at all. It is represented by the symbols Ω or S. On the other hand, the impossible event is an event that cannot happen and is shown as an empty set ($\emptyset$).

We group events into different categories to help us focus on specific situations or results that we are interested in. This way, we can study how likely these events are and use this information to make smart choices (Shiryaev, 2016).

Knowing the ideas of sample space and events is very important for measuring and understanding uncertainty. By determining the different outcomes that can happen, we set the range of possible results. And by specifying the specific combinations or smaller groups of results that we want to pay attention to, we can narrow down our focus. These ideas are the starting point for studying probability theory. They help us figure out how likely events are, make predictions, and make good decisions when we can't be certain about the outcome of random experiments (Epp, 2010).

9.2.2. Probability Measure

A probability measure is an important idea in probability theory that calculates how likely events are to happen within a certain group of possibilities. It helps us assign numbers to events to show how likely they are to happen. A probability measure is a way to assign numbers to events. These numbers are non-negative and represent the likelihood of each event happening. It meets several important characteristics. First, the chance of something happening is always positive, meaning it cannot be negative or less than zero. This feature makes sure that probabilities are always positive real numbers.

The additivity rule says that if two events, A and B, cannot both happen at the same time, then the chance of either event happening (or both) is equal to the chance of event A happening plus the chance of event B happening. This idea also applies to adding up groups of separate events that don't overlap. It helps us figure out the chance of complicated events by looking at the chance of their different parts happening. Additionally, the total probability property says that for any set of possible outcomes, the chance of all the possible outcomes happening is 1. This means that there is an event in the sample space that is guaranteed to happen. It makes sure that the total probability of all possible outcomes is 1 (Shiryaev, 2016).

Complementary events are very important when we calculate probability. The chances of something not happening, which is called the complement of event A, is equal to 1 minus the chances of it happening. Complementary events happen when the opposite of event A occurs. This feature helps us figure out how likely it is that something won't happen by taking away its likelihood of happening from 1. Assigning probabilities to events in the sample space means figuring out how likely each event is compared to the others. You can solve this through careful observation, mathematical models, or thinking clearly, depending on the situation. When there are many different things that could happen and they all have the same chance of happening, it's easier to assign probabilities to them. Imagine you have a regular dice with six sides. When you pick it up and turn it around, each side has the same likelihood of ending up at the top. This means that there is an equal chance of 1 out of 6 for each side to happen (Epp, 2010). In more complex situations, advanced methods are used to figure out probabilities. Probability distributions are math tools that help us figure out how likely something is to happen in certain situations. There are different types of probability distributions like the binomial distribution or the normal distribution. They help us make accurate estimates of probabilities in specific situations. Probability measures are used in many different areas. Risk assessment is really important because it helps people and groups figure out and control risks. Probability measures are important for statistical inference. They help us draw valid conclusions using sample data. Probabilities are important in decision-making because they help us assess different outcomes and make the best choices. Moreover, probability measures are important in predictive modeling and forecasting because they help us measure uncertainty and make correct predictions about what will happen in the future.

Basically, probability measures are really important in probability theory. They give a way to measure how unsure we are and assign numbers to the likelihood of different events happening. Probability measures have many uses, such as determining risks and making predictions for decision-making. By using probability measures, we can figure out how likely things are to happen. This helps us make better decisions and deal with uncertainty in everyday situations (Shiryaev, 2016).

9.2.3. Set Operations and Probability Laws

Set operations are helpful tools for organizing things and making new things by understanding how they are connected in a group of choices. Probability theory uses three main ways to combine groups: joining them (union), finding their similarities (intersection), and identifying what is not in them (complement). The union of two events A and B, written as A ∪ B, means that either A or B, or both, happen. This means it includes all the things in A and B, or in both of them. When we say A ∩ B, it means that both A and B happen together. Both A and B have the same outcomes. The complement of an event A means that event A did not happen. This means there are some results that do not belong to group A but are still considered as possible outcomes. Complementary events are when the results are completely different from each other. These actions can help us put together, compare, and analyze events so that we can think about different options and make smart choices (Epp, 2010).

Probability laws are important rules that control how probabilities work and help us figure out how to calculate probabilities. The addition law of probability, also called additivity, says that when you combine two events A and B, the chance of both happening together is equal to the chance of each event happening on its own, minus the chance of both events happening at the same time. This law helps us figure out how likely complex events are by looking at how likely their smaller parts are. In simple terms, the multiplication law of probability says that if two events A and B are independent, the chance of both of them happening together (the intersection) is equal to the chance of event A happening multiplied by the chance of event B happening. This law is used when one event happening does not change the chance of another event happening. It shows that the events are not related to each other. The complement law says that if you want to know the chance of something not happening, you can subtract the chance of it happening from 1. This rule lets us find out how likely it is for something not to happen by taking away its likelihood from 1 (Shiryaev, 2016) (Figure 9.2).

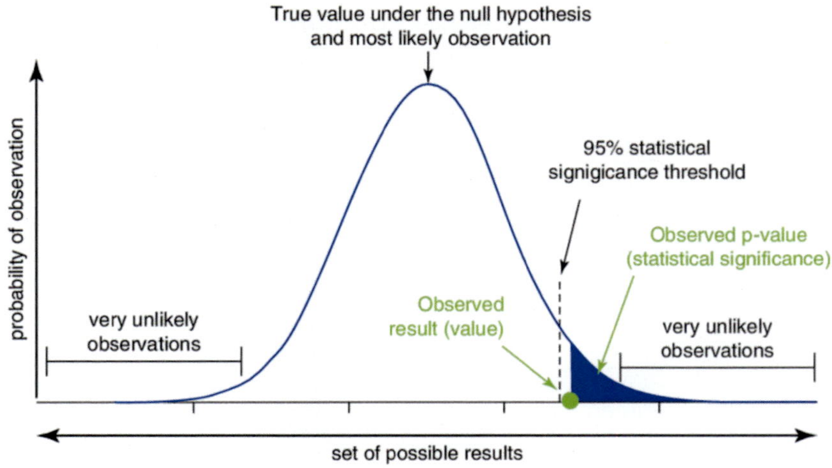

Figure 9.2. The normal law of probability distribution.

Source: Nundy, Kakar, & Bhutta (2021); Scientific Figure on ResearchGate.

These actions of combining sets and rules of probability are useful tools for working with events, figuring out chances, and thinking about situations where the outcome is unsure. By using these ideas, we can break down difficult situations, figure out how likely certain things will happen, and evaluate the chances of different results. These basic ideas help us measure uncertainty, make smart decisions, and use probability theory well in different areas like statistics, decision-making, and evaluating risks. When we understand how sets work and the rules of probability, we can use probability theory to help us analyze and measure uncertainty in different real-life situations (Epp, 2010).

9.2.3.1. Complementary Events and Odds

In probability theory, complementary events and odds are ideas that help us understand how different events are related and offer different ways to think about probabilities. They give us information about what could happen in a situation and offer another way to talk about how likely something is, so we can think about not being sure about things from different perspectives (Shiryaev, 2016).

Complementary events are two events that show opposite results in a group of possibilities. To put it simply, the complement of an event A means that A does not happen. It includes results that are not in group A but are still within the possibilities. Complementary events are helpful when looking at different outcomes in probability analysis. By looking at both an event and its opposite, we can cover all the different results that could happen.

Complementary events are connected by a rule called the complement law. This rule says that the probability of the opposite of an event A is equal to one minus the probability of A: P(A') = 1 – P(A). This law helps us figure out how likely it is for something to not happen by taking away its likelihood from 1. Complementary events help us see probabilities from a different angle. They let us think about what doesn't happen in order to understand uncertainty.

Another similar idea is called odds, and it's a different way to talk about probabilities. The chances of something happening, called Odds(A), is found by dividing the probability of it happening by the probability of it not happening. So, Odds(A) = P(A)/P(A'). Chances are usually shown as a ratio or as a fraction. For instance, if there's a 60% chance that event A will happen, then there's a 40% chance that event A will not happen. So, the chance of A happening is 3/2. Odds give us a different way to think about chances by comparing how likely something is to happen with how likely it is not to happen. They help us determine how likely different results are and compare different events. For example, if event A is more likely to happen than event B, it means that event A is more likely to occur (Shiryaev, 2016).

Complementary events and odds help us understand how events relate to each other and offer different ways to think about and describe probabilities. We can understand all possible outcomes by looking at both an event and its opposite. The odds help us see how likely different things are and compare them to each other. These ideas help us to think better about situations where we are not sure what will happen, and to make smart choices considering different points of view (Epp, 2010).

To put it simply, complementary events and odds help us understand probabilities better in probability theory. Complementary events are when we consider two outcomes that are opposite from each other in a group of possible outcomes. This helps us think about all the different possibilities. The complement law allows us to figure out the chance of something not happening. Odds are a different way to show how likely something is to happen compared to not happening. These ideas make our analysis of chances better and help us understand uncertainty and decision-making more deeply.

9.2.4. Conditional Probability

In probability theory, conditional probability means figuring out the likelihood of one thing happening after another thing has already happened. It helps to modify probabilities when we get new information or conditions. Conditional probability is important for studying events that are related to each other, and how the chance of one event happening can affect the chance of another event happening. In simpler terms, conditional probability is a way to find the likelihood of event A happening, knowing that event B has already happened. It is written as P (A | B). The line "|" means "given" or "conditional on. " To find the probability of A given B, use the formula: P (A | B) = P (A ∩ B)/P(B), where P (A ∩ B) is the probability of both A and B happening at the same time, and P(B) is the probability of event B happening (Shiryaev, 2016).

Conditional probability is about updating our knowledge and changing probabilities when we learn something new. Let's think about a pack of cards as an example. If we pick one card from a deck and it is shown to be a heart, the chances of picking another heart from the rest of the cards are different. At first, there was a 13/52 chance of getting a heart, but when the first heart was seen, there were now only 51 cards left with 12 hearts. So, if the first card is a heart, the chance of drawing another heart is 12 out of 51 (Epp, 2010).

Conditional probability is very important when things are connected, meaning that one thing happening can change the chance of another thing happening. It helps us look at situations where what happens is affected by what we already know or by what is happening around us. For example, in medical testing, conditional probability is used to figure out how likely it is for someone to have a specific disease based on their symptoms or test results.

Bayes' Theorem is a helpful rule in math that helps us calculate probabilities based on other probabilities. This formula proves that if event B has happened, the probability of event A happening is the same as the probability of event B happening if event A has already happened, times the probability of event A happening, divided by the probability of event B happening. Bayes' Theorem is frequently used in subjects like math, computer science, and investigating information (Shiryaev, 2016).

Conditional probability helps us when we are unsure about something, by providing a useful way to think about different situations. It helps us change our opportunities when we learn new things or situations. It helps us predict how likely something will happen based on the situation or what happened before. By using conditional probability when analyzing data, we can get better at making accurate predictions, making smarter choices, and understanding how different events are connected to one another.

In simple terms, conditional probability is a concept in probability theory that helps us determine the probability of one event happening, based on another event happening first. By using conditional probability in our analysis, we can better understand uncertainty and make smart decisions in many areas like medicine, finance, data analysis, and machine learning (Epp, 2010).

9.2.5. Independence and Dependence

Independence and dependence are ideas in probability theory that explain how events are related to each other. Independence means that one event happening doesn't affect the chances of another event happening. Dependence means that when one event happens, it can make another event more likely to happen too. It is important to know about independence and dependence to understand uncertainty in different fields like statistics, finance, and machine learning. These ideas help us understand how multiple things happen and allow us to make good guesses and smart choices.

9.3. COUNTING PRINCIPLES AND PROBABILITY

Counting principles are the foundation for figuring out chances and measuring uncertainty. By knowing how to count, we can figure out the chances of something happening and make smart choices. The Multiplication Principle is a rule that says when we want to find the total number of outcomes for a series of events, we can multiply the number of outcomes for each event together. The multiplication principle says that if you can do one thing in n1 ways and another thing in n2 ways, then you can do both things one after the other in n1 × n2 ways. This rule applies when the tasks don't depend on each other. By using the multiplication principle, we can figure out the total number of outcomes for a sequence of separate events. For instance, let's think about tossing a coin and rolling a dice. The coin can either have a head or a tail, while the die can show any number from 1 to 6. By using the multiplication principle, we can figure out that when we flip a coin and roll a die at the same time, there are 12 different things that could happen altogether (Shiryaev, 2016).

The addition principle says that if you can do the first task in n1 different ways and the second task in n2 different ways, then you can do either task in a total of n1 + n2 ways. This rule applies when the tasks cannot happen at the same time. By using the addition principle, we can find out how many outcomes there are when we have a bunch of events that can't happen at the same time. For example, imagine you are picking a card from a regular deck of 52 cards. If we want to find out how many different ways, we can draw a heart or a spade, we can use the addition principle. In a deck, there are 13 hearts and 13 spades. If you draw either a heart or a spade, there are 26 different possibilities (Epp, 2010).

Permutations and combinations are ways to count when the order or the choice of things matters. Permutations mean putting objects in a certain order. To find the number of ways to arrange objects, we use formulas like $nPr = n./(n - r)$. In this formula, n is the total number of objects and r is the number of objects being chosen. To find the number of combinations, we can use a formula called $nCr = n./(r \times (n - r)!)$. Here, 'n' stands for the total number of objects and 'r' stands for the number of objects we choose.

These rules help us figure out how many different results could happen, and then we can use that information to find the chances of each outcome. To find the probability, you divide the number of good outcomes by the total number of possible outcomes. By using counting principles, we can accurately measure the chances of something happening and measure how uncertain it is. It is important to understand how to count things correctly for many different uses of probability theory. This includes statistics, figuring out combinations, and creating algorithms. These principles help us count and calculate possibilities and make smart choices using the chance of something happening. By using counting principles, we can deal with uncertainty, study chances, and improve our understanding of the world (Shiryaev, 2016).

9.4. DISCRETE RANDOM VARIABLES

A random variable is a number that indicates the outcome of something that happens by chance in probability and statistics. Any value, not just a specific one, can exist for a discrete random variable. It implies that you can calculate the likelihood of each potential outcome in a random circumstance. A discrete random variable is a type of variable that can only have specific values that are simple to count. It can have certain numbers, like whole numbers, with certain chances given to each number. For instance, when you flip a coin, the number of times it lands on heads is a random whole number that you can't predict. Put simply, if you keep track of how many cars go through a junction or the number of mistakes made in a production line, these are also instances of things that can change unexpectedly.

The probability mass function (PMF) helps us know how likely different X values are to occur. The PMF can tell us the probability of something happening. The PMF has two important rules: Every value has a chance that is not negative, so the chance of X being the same as x is always equal to or greater than 0 for any value of x. The total chance of all possible values is equal to 1, so when we add up the chances of X being the same as x for all possible values of x, the total should be equal to 1 (Epp, 2010).

The PMF gives all the information about the chances of different outcomes of a variable. It helps us figure out the chances of certain results and make predictions about the random thing we're studying. By studying the PMF, we can figure out how likely different values are and get a sense of how the random variable is spread out. The cumulative distribution function (CDF) of a discrete random variable X shows the chance that X is equal to or less than a specific value. The CDF, known as F(x), gives us details about the probabilities that add up to a certain value or less. It helps to find the percentiles and cumulative probabilities (Shiryaev, 2016).

The average or mean of a random variable X tells us where the majority of its values tend to be. It shows the expected average value of X over a long period of time. To find the expected value of X, we add up all the possible values of X together with their respective probabilities. The variation or range of different values that a random number X can have, in relation to the average value, is measured as variance. We measure how much the values differ from the average. Understanding the average value and spread of a random number that can only take certain values helps us understand how it is spread out and make informed guesses and analyze information. By doing some calculations, we can figure out the average result and see how much the results are spread out. Some examples of probability distributions for situations where things can only have certain values are the binomial distribution and the Poisson distribution. Every distribution has its own PMF and certain qualities that assist in comprehending and studying the random variable. Basically, a discrete random variable is when there are a limited number of possible outcomes, and each outcome has a certain probability. The PMF gives probabilities to different values and helps us find the probabilities of certain results. The expected value and variance help us understand the average and spread of a random variable. Probability distributions help to explain how discrete random variables behave. It is important to understand these ideas to do statistical analysis, model real-life things, and make informed choices using probabilities.

9.5. COMMON DISCRETE PROBABILITY DISTRIBUTIONS

In mathematics and statistics, different ways are used to represent and study things that can happen randomly and have only a specific set of values. These distributions help us understand how random things work and figure out the chances of different results happening. Let's learn about some basic probability patterns that can happen one at a time (Shiryaev, 2016).

- The Bernoulli Distribution is a pattern that shows the probability of getting a certain outcome (such as heads or tails) in a single trial or event. It is based on the concept of a binary outcome, where there are only two possible results. The Bernoulli distribution is used to describe an experiment that has only two possible outcomes: either a success or a failure. These outcomes are usually represented by the numbers 1 and 0. It is described by one number, p, which shows the likelihood of something going well. The chance of an event in the Bernoulli distribution happening is described by a function called the PMF. The PMF formula is $P (X = x) = p^x \times (1 - p) ^ (1 - x)$. x can only be either 0 or 1 (Epp, 2010).
- The binomial distribution is a mathematical model that is used to describe the probability of a certain number of successes in a fixed number of trials, where each trial has only two possible outcomes, usually referred to as "success" and "failure." The binomial distribution is commonly used to represent the number of times something succeeds in a certain number of separate attempts. This distribution is described by two things: how many tries you do (n) and how likely it is to succeed (p). The probability of getting exactly k successes in a binomial experiment with n trials is given by the binomial distribution. The formula for this probability is $P (X = k) = C (n, k) \times p^k \times (1 - p) ^ (n - k)$, where k is the number of successes and C (n, k) is the number of ways to choose k objects out of n.
- The Poisson Distribution is a mathematical concept that helps us understand the probability of a certain number of events occurring

in a fixed interval of time or space. It is often used to study rare events or situations where the average rate of occurrence is known, but the specific timing or number of events is uncertain. The Poisson distribution is often used to show how many things happen within a certain timeframe or area, when those things don't happen very often and don't depend on each other. It is described by one number, called lambda, that shows how often it happens on average. The PMF of the Poisson distribution is a formula that calculates the probability of getting a certain number of events in a certain time period. The formula is $P(X = k) = (e^{\wedge}(-\lambda) \times \lambda^{\wedge}k)/k.$, where k is the number of events.

- The geometric distribution is used to count how many attempts it takes to achieve the first success in a series of individual attempts. A value known as "p" tells us how likely it is that each attempt will succeed, and it serves as the basis for this decision. By adding (k minus 1), (1 minus the probability of success), and the probability of success, it is possible to determine the likelihood of success after k attempts (Shiryaev, 2016).
- The negative binomial distribution helps us understand how likely it is to have a certain number of failures before achieving a certain number of successes in a series of independent trials. The negative binomial distribution helps us figure out how many attempts it takes to achieve a specific number of successes in a series of experiments where the answer can only be yes or no. It is like the geometric distribution. This can be understood by two things: how often something is successful (r), and how likely it is to be successful (p). The PMF of the negative binomial distribution tells you how likely it is to get a certain number of successes, denoted as "X," in a given number of attempts. The formula to calculate the probability is $P(X = k) = C(k - 1, r - 1) \times p^{\wedge}r \times (1 - p)^{\wedge}(k - r)$. In this equation, the letter "k" stands for the total amount of experiments or tries (Epp, 2010).

If we understand these patterns, we can use them to represent and study different things that happen in the real world. We can also guess what might happen in the future and figure out how likely something is based on the information we have. Probability distributions are really important in statistics and have a big impact on making decisions, assessing risk, and doing scientific research.

9.6. LAW OF LARGE NUMBERS AND CENTRAL LIMIT THEOREM

The Law of Large Numbers is an important idea in probability theory and statistics. It explains how sample averages change when we have more things in our sample. It means that if we have more observations or trials, the average of those observations gets closer to the expected value or average value of the whole group. In simpler words, this means that if we have more information, our guesses or calculations will be more precise (Shiryaev, 2016) (Figure 9.3).

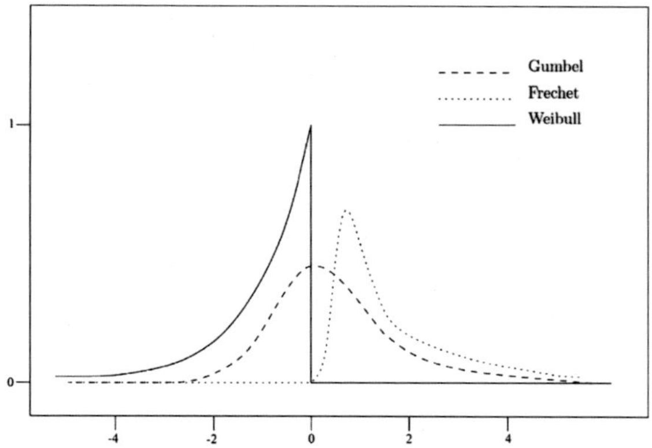

Figure 9.3. Densities of the standard extreme value distribution.

Source: Ďuriš, Bartková, & Tirpáková (2021).

The Law of Large Numbers has two main forms: the Weak Law of Large Numbers and the Strong Law of Large Numbers. The Weak Law of Large Numbers says that as we gather more and more samples, the average value of those samples gets closer to what we expect it to be. In simpler terms, as we gather more data, the average of the observations gets closer to the actual average. This law makes sure that, typically, our guesses are correct and trustworthy. The Strong Law of Large Numbers says that the average of a sample will almost always get closer to the expected value. This means that as the sample size gets bigger, there is a very low chance that the average of the sample will differ a lot from the actual mean value. Put simply, if you have a big enough sample, the average of the sample will be very close to the actual average of the whole population.

The Law of Large Numbers is very important in different areas like making decisions, assessing risks, and making conclusions based on statistics. This helps us make accurate predictions and draw conclusions from data. By knowing this law, when we have more samples, we can trust that the average of those samples will stay the same and be reliable. This helps us make better conclusions and decisions using statistics (Shiryaev, 2016). The Central Limit Theorem is an important idea in probability and statistics that talks about what happens when you add or average many random numbers that are all the same. This statement is saying that no matter what the original distribution looks like, the distribution of the total or average of a sample will start to look like a bell-shaped curve called the normal distribution, or Gaussian distribution, as the sample size gets bigger.

The Central Limit Theorem has some really important implications:

1. **Normal Approximation:** We can estimate the pattern of a group's total or average by using the normal distribution, even if the individual numbers are not typically distributed. This estimation is useful because the normal distribution has been well researched and has many properties that make statistical analysis easier (Epp, 2010).
2. **Confidence Intervals:** These are important because they give us a range of likely values for certain characteristics of a population. The Central Limit Theorem is used to help figure out these intervals. Confidence intervals are commonly used in statistics to estimate things like the average or proportion of a whole population. They help give an idea of how certain we can be about our estimates.
3. **Hypothesis Testing:** This theorem helps us test ideas by looking at some information we collected. It helps us figure out how likely a belief or claim is to be true. This helps us figure out test numbers like z-scores, which tell us how significant differences are and if our ideas are valid.
4. **Central Limit Theorem:** Determining the right sample size for statistical studies often involves using the Central Limit Theorem. This tool will help you figure out how many samples you need to get accurate results in your statistical estimates (Shiryaev, 2016).

The Central Limit Theorem is a helpful tool in statistics. It supports many statistical methods and helps us draw strong conclusions from data. Its ability to be used in many different situations and its trustworthiness make it a very important part of statistics. By using the theorem, researchers and analysts can learn important information, make correct guesses, and reach valid conclusions even in situations where the individual observations don't follow a usual pattern.

9.7. APPLICATIONS OF DISCRETE PROBABILITY

Discrete probability is used in many different areas. In finance, it uses math to predict how stock prices will change and to measure the amount of risk in investment plans. In computer science, it is the basis for processes used in analyzing data, teaching machines, and protecting information. In genetics, it helps us to understand how traits are passed on from one generation to the next and to make predictions about how genes will affect different characteristics. In the study of disease, it helps us understand how diseases spread and evaluate if intervention measures are effective. In operations research, it helps improve efficiency and decision-making in managing and organizing logistics and supply chains. In gaming and gambling, it helps you figure out the best strategies and chances of winning. Discrete probability helps us use math to study uncertainty, make guesses about the future, and help us make decisions in various areas of life (Shiryaev, 2016).

To sum up, discrete probability is an important idea in math that helps us understand and measure uncertainty in different real-life situations. It gives you helpful tools for figuring out the chances of something happening, understanding how things that happen by chance work, and making smart choices based on information. Discrete probability is really important in fields like statistics, finance, computer science, and other areas (Epp, 2010).

CHAPTER 10

Recurrence Relations

Contents

10.1. INTRODUCTION

In this section, we will begin learning about recurrence relations and their solutions. Our major focus will be on the class of finite order linear recurrence relations with constant coefficients, also known as finite order linear relations. We'll start with the closed form expressions that give rise to these relations. Then we'll provide you an algorithm to solve them (Miller, 2011).

The concept of recurrence relations, which requires establishing a sequence of numbers or values based on terms from the sequence's past, is a crucial idea in mathematics. They are especially useful when addressing repeated processes or occurrences when the present value is dependent on one or more preceding values.

Each phrase in a recurrence relation is decided by a rule or formula that takes preceding terms into account. This rule frequently specifies how the present term links to one or more terms in the sequence that came before it. Knowing the beginning terms and the recurrence relation allows us to generate the next terms in the series.

There are two kinds of recurrence relations: linear and nonlinear. As their current term, simple linear recurrence relations have a current term that is a linear mixture of preceding terms multiplied by a constant. Nonlinear recurrence relations, on the other hand, necessitate more complex interactions between terms and commonly include exponential, logarithmic, or trigonometric functions.

To solve recurrence relations, you must find a closed-form expression or formula that computes the nth term of the series without consulting any previous terms. This approach commonly employs a variety of methods, including substitution, iteration, characteristic equations, generating functions, and the solution of linear equation systems. Solving recurrence relations provides informative information about the behavior and properties of sequences, in addition to assisting in the analysis of algorithms, population models, and other dynamic systems.

Recurrence relations are frequently employed in many branches of mathematics as well as in other fields. They are used to define dynamic systems in physics, to measure the temporal complexity of algorithms in computer science, to count objects in combinatorics, and to solve differential equations in calculus. They are also important in fields such as engineering, biology, economics, and many others where the evolution of a system is dependent on its preceding states, such as those (Miller, 2011).

10.2. DEFINITION AND TERMINOLOGY

Consider S to be a list of numbers. With the exception of a tiny number, all terms of S are linked by a formula known as a recurrence relation on S. That is, there is a k_0 in the domain of S such as if $k > k_0$, then S(k) can be expressed in terms of some of the terms that precede 5(k). If the domain of S is {0, 1, 2,}, the terms 5(0), 5(1), S (k_0–1) are not defined by the recurrence formula. Their values are the initial conditions (or basis or boundary conditions) that complete the definition of S.

10.2.1. Rabbits and the Fibonacci Numbers

As an example, suppose a juvenile pair of bunnies—one of each gender—is abandoned on an island. When the bunnies are two months old, they begin to reproduce. After the age of two months, each pair of rabbits produces a new pair every month. Determine a recurrence relation for the number of pairs of rabbits on the island after n months, assuming rabbits never die. In the 13th century, Leonardo Pisano (Fibonacci) examined this subject for the first time. Solution: Let fn be the number of rabbits after n months. We'll show that the Fibonacci sequence's terms are fn, where n = 1, 2, 3, … Due to the fact that this couple does not reproduce during the second month, the number of rabbit pairs on the island at the end of the first month is f1 = f2 = The number of pairs after n months = "old" pairs + new pairs.

10.2.2. The Tower of Hanoi

In the late 1800s, French mathematician Édouard Lucas invented a puzzle with three pegs on a board with varying-sized disks. The largest disk is placed at the bottom of the first peg, followed by the other disks in order of size (Miller, 2011).

You may move the disks from one peg to another one at a time as long as a larger disk is never placed on a smaller disk.

The goal is to position the disks on the second peg in increasing size order, with the largest disk at the bottom, after using all authorized moves. Assume that Hn is required to use n disks in order to accomplish the Tower of Hanoi puzzle.

The puzzle played a role in the construction of a myth. Monks in a Hanoi tower are shifting 64 gold disks from one peg to another in line with the puzzle's instructions.

They rotate one disk every day. When the puzzle is finished, the world will come to an end.

Using this formula, it takes $2^{64} - 1$ = 18,446, 744,073, 709,551,615 days, or more than 500 billion years, to solve the riddle for the 64 gold disks in the story.

The Reve's puzzle, proposed by Henry Dudeney in 1907, is similar but has four pegs. A well-known unsolved conjecture is the least number of movements required to complete this puzzle.

10.2.3. Counting Bit Strings

Determine a recurrence relation for the number of bit strings of length n that do not contain two consecutive 0s, then explain the initial conditions. How many such five-bit string bit strings exist?

Solution: a_n is the number of n-bit strings that do not contain two consecutive 0s.

Example: Some examples of recurrence relations:

I. The Fibonacci sequence was defined by the recurrence relation Fk = Fk_ **2** + Fk_i, k > 2, with the initial conditions F_0 = 1 and F_1 = 1. The recurrence relation is called a second-order relation because Fk depends on the two previous terms of F.

II. The relation $T(k) = 2T(k - 1)^2 - kT(k - 3)$ is a third-order recurrence relation. If values of T(0), T(1), and T(2) are specified, then T is completely defined.

III. The recurrence relation 5(n) = S([n/2 i) + 5, n > 0, with 5(0) = 0 has infinite order. To determine 5(n) when n is even, you must go back n/2 terms. Since n/2 grows unbounded with n, no finite order can be given to S.

10.3. SOLVING RECURRENCE RELATIONS

Although recurrence relations are typically the simplest approach to express sequences, they can be time-consuming to employ when directly computing terms. The process of generating a closed form expression for the terms of a sequence from its recurrence relation is known as solving the relation. All recurrence relations cannot be solved using a single method or algorithm. Some recurrence relations are, in fact, unsolvable. One such example is the above-mentioned relation. The most common sort of recurrence relations you will encounter in the future are finite order linear recurrence relations with constant coefficients. This lesson will take up the most of our time in this chapter (Miller, 2011).

Definition 10.3.1 nth Order Linear Recurrence Relation

Let S be a list of numbers with domain k > 0. An nth order linear recurrence relation on S with constant coefficients is a recurrence relation that can be written in the form.

S(k)± CiS(k –1)± …+C…S(k – n) = f(k) for k > n where C1, C2 • • Cn are constants and I is a numeric function that is defined for k > n.

Note: We will shorten the name of this class of relations to n^{th} order linear relations. Therefore, in further discussions, 5(k) ±2kS(k – 1) = 0 would not be considered a first-order linear relation.

Example: Some Finite Order Linear Relations

I. The Fibonacci list is defined by the second-order linear relation because F1, –Fk_i – Fk_2 = 0
II. The relation P(j) 2P(j – 3) = j^2 is a third-order linear relation_ In this case, C_1 = C2 = 0.
III. The relation A(k) = 2(A(k – 1) + k) written as A(k) – 2A(k – 1) = 2k. Thus, it is a first-order linear relation.

10.4. RECURRENCE RELATIONS OBTAINED FROM "SOLUTIONS"

Prior to giving an algorithm for solving finite order linear relations, examine recurrence relations that can arise from some closed form expressions. This closed form expressions are selected so as one will obtain finite order linear relations from them. The approach can be seen as a bit contrived, but if one were to write down a few algebraic expressions, chance is that most of them would be identical to the ones that are about to be examined.

For our first example, consider D, defined by D(k) = 5.

2^k, k > 0.

If k> 1, D(k) = 5.

2^k = 2.5

2^{k-1} = 2D(k – 1).

Therefore, D satisfies the first order linear relation D(k) – 2D(k – 1) = 0 and the initial condition D(0) = 5 serves as an initial condition for D (Miller, 2011). As a second example, consider C(k) = 3k–1 + $_2$+1 $_{k\,k}$0. Quite more algebraic manipulation is required to get our result: The recurrence relation that we have just obtained, defined for **k** > 2, together with the initial conditions C(0) = 7/3 and C(1) = 6, define C. Based on these results, we might conjecture that any closed form expression for a sequence that combines exponential expressions and polynomial expressions will be solutions of finite order linear relations. Not only is this true, but the converse is true: a finite order linear relation defines a closed form expression that is similar to the ones that were just examined. The only additional information that is needed is a set of initial conditions.

10.4.1. Recurrence Relations Obtained From Given Sequences

Recurrence relation:

D(k) – 2D(k – 1) =

C(k) – 2C(k – 1) – 6C(k – 2) = 2k – 7

Q(k) – Q(k –1) = 2

A(k) – 2A(k – 1) + – 2) = 2

B(k) – 2.13(k – 1) + B(k – 2) =4

G(k) – G(k – 1) + 12G(k – 2) =0

J(k) – 4.1(k – 1) + 4J(k – 2) =0

Closed form expression:

D(k) = 5 • 2/’

C(k) = ak-i 2k-i   k

Q(k) = 2k + 9

A(k) = k2 – k

13(k) = 2k2 + 1

G(k) = 2 – 5(_3)k

J(k) = (3 + k)2k

10.5. HOMOGENEOUS RECURRENCE RELATION

An nth order linear relation is homogeneous if f (k) = 0 for all k. For each recurrence relation S(k) + Ci. 5 (k – 1) ±… + C_7,5 (k – n) = f(k), the associated homogeneous relation is 5(k) + CiS(k – 1) ±… + C.S(k – n) =0

Example: First order homogeneous recurrence relations

D(k) – 2D(k – 1) = 0 is a first-order homogeneous relation. Since it can also be written as D(k) = 2D(k – 1), it shouldn't surprise that it arose from an expression that involves powers of 2. Generally, one would expect that the solution of L(k) – aL(k – 1) would involve ak. Actually, the solution is $L(k) = (0)a^k$, where the value of L(0) is given by the initial condition (Miller, 2011)

Example: A second order example

Consider the second-order homogeneous relation S(k) – 75(k – 1) ± 125(k – 2) = 0 together with the initial conditions 5(0) = 4 and 5(1) = 4. From the discussion above, one can predict that the solution to this relation involves terms of the form ba^k, where b and a are nonzero constants that must be determined. If the solution were to equal this quantity exactly, then,

5(k) = bak

$5(k-1) = ba^{k-1}$

$5(k-2) = ba^{k-2}$

Substitute these expressions into the recurrence relation to get:

$Ba^{k1} - Ma^{k-I} + 12ba^{k-2} = 0$

Each term on the left-hand side of the equation has a factor of ba^{k-2}, which is nonzero. Dividing through by this common factor yields

$a^2 - 7a + 12 = (a-3)(a-4) = 0$ (10.3.1)

Thus, the only possible values of a are 3 and 4. Equation (10.3.1) is known as the characteristic equation of the recurrence relation. The fact is that the original recurrence relation is true for any sequence of the form 5(k) = $b_1 3k\ b_2 4^k$, where b_1 and b_2 are real numbers.

This set of sequences is known as the general solution of the recurrence relation. If one didn't have initial conditions for 5, it would stop here. The initial conditions make it possible for one to find definite values for b_1 and the solution of this set of simultaneous equations is $b_1 = 12$ and $b_2 = -8$ and so the solution is $5(k) = 12\ 3^k - 8 \cdot 4^k$.

f bi3n b24° = 4 b2 = 4

b131 +b241 =4 3b1 + 4b2 = 4

Definition 10.5.1: Characteristic Equation

The characteristic equation of the homogeneous nth order linear relation

S(k) + C1 S(k – 1) + C„5 (k – = 0 is

the nth degree polynomial equation

$+ E = \pm Cia^{n-1} \pm C_{n.} = 0$

i=1

The left-hand side of this equation is called the characteristic polynomial. The roots of the characteristic polynomial are called the characteristic roots of the equation (Miller, 2011).

Example: Some Characteristic Equations

i. The characteristic equation of F(k) – F(k – 1) – F(k – 2) = 0 is $a^2 - a - 1 = 0$.

ii. The characteristic equation of Q(k) + 2Q(k – 1) – 3Q (k – 2) – 6Q (k – 4) = 0 is $a^4 + 2a^3 - 3a^2 - 6 = 0$.

Solve the following sets of recurrence relations and initial conditions:

1) 5(k) – 105(k – 1) + 95(k–2) = 0, 5(0) = 3, 5(1) =11

Answer:

5(k) = 2 + 9k

2) 5(k) – 95(k – 1) + 185(k– 2) =0, 5(0) = 0, 5(1) = 3

Note that the absence of a Q(k – 3) term means that there is not an x^{4-3} = x term appearing in the characteristic equation.

10.5.1. Algorithm for Solving Homogeneous Finite Order Linear Relations

i. Write out the characteristic equation of the relation

   $5(k) + C_i$ S(k – 1) C_7,5(k – n) = 0, which is $_{ar}t \pm$ ce2 – 1 ± $C_{tli}a$ C_n. = 0.

ii. Find all roots of the characteristic equation, the characteristic roots.

iii. If there are n distinct characteristic roots, a_1, a_2, … a, then the general solution of the recurrence relation is 5(k) = biaik b2a2^k + • • • + $b_n a_n^k$. If there are fewer than n characteristic roots, then at least one root is a multiple root.

If a_i is a double root, then the bjaj^k term is replaced with $(b_j k)a$. In general, if a_i is a root of multiplicity p, then the bjajk term is replaced with (b_0 b31 k + • • • + b j ^{1}) a^3.

iv. If n initial conditions are given, we get n linear equations in n unknowns (the Vs from Step 3) by substitution. If possible, solve these equations to determine a final form for 5(k).

Although this algorithm is valid for all values of n, there are limits to the size of n for which the algorithm is feasible. Using just a pencil and paper, we can always solve second-order equations. The quadratic formula for the roots of ax^e bx C =

is

–b ± b^2 – 4cie x =

2a

The solutions of a^2 Ci a + C2 = 0 are then 21'(C_1^2 – 4C2) and –(C1^{24}C2)

Although cubic and quartic formulas exist, they are too long to introduce here. Thus, the only higher-order relations (n > 3) that one could be expected to solve by hand are ones for which there is an easy factorization of the characteristic polynomial (Miller, 2011).

Example: A solution using the algorithm

Suppose that T is defined by T(k) = 77^1(k – 1) – 10T(k – 2), with T(0) = 4 and T(1) = 17. One can solve this recurrence relation with Algorithm 10.3.1:

i. Note that recurrence relation is written in "nonstandard" form. To avoid errors in this step, one can consider a rearrangement of the equation to,

T(k) – 77^1(k – 1) ± 10T(k – 2) = 0.

Thus, the characteristic equation is $a^2 - 7a + 10 = 0$.

ii. The characteristic roots are

(7 +/49 – 40) = 5 and (7 – N/49 – 40) = 2.

These roots can be just as easily obtained by factoring the characteristic polynomial into (a – 5) (a – 2).

iii. The general solution of the recurrence relation is

T(k) = ba^k b25^k •{T(0) = 4 1 $b_i 2°$ $b_2 5°$ = 4 fb$_i$ + b_2 = 4

iv. The simultaneous equations have the solution

T(1) = 17 t bi2^1 b25^1 = 17t 2b_i 5b_2 = 17 b_i = 1 and b_2 = 3.

Therefore, T(k) = 2^k ± 3 • 5^k.

The following is one such guideline: If the coefficients of the characteristic polynomial are all integers and the constant term is equal to m, the only rational characteristic roots that are viable are divisors of in (both positive and negative). We can use a computer (or perhaps simply a calculator) to calculate n. A number of well-known approaches, some of which are included in mainstream software packages, can be used to estimate the characteristic roots. There is no general rule that specifies the values of n for which numerical approximations are possible. The accuracy you acquire will be determined by the relationship you attempt to solve (Miller, 2011).

Example: Solution of a third order recurrence relation

Solve S(k) – 75(k – 2) + 65(k – 3) = 0, where S(0) = 8, 5(1) = 6, and S(2) = 22.

i. The characteristic equation is a – 7a + 6 = 0.

ii. The only rational roots that we can attempt are ±1, ±2, and ± 6. By checking these, we obtain the three roots 1, 2, and –3.

iii. The general solution is 5(k) = $b_1$1k b22^k b3($^-$3)^k. The first term can simply be written b1.

a. 5(0) = 8

bi b2 = 8

iv. 5(1) = 62b2 – 3b3 = 6

5(2) = 22

4b_2 9b_3 = 22 b_3= 1.

Therefore, 5(k) = 5 + 2 2^k + (–3)^k = 5 + 2k+^{1} –+ (–3)^k

Example: Solution with a double characteristic root

Solve D(k) – 8D(k – 1) + 16 D(k – = 0, where D(2) = 16 and D(3) = 80.

i. Characteristic equation: a^2 – 8a + 16 = 0.

ii. a^2 – Set + 16 = $(a – 4)^2$. Therefore, there is a double characteristic root, 4.

iii. General solution: D(k) = (b10 bilk) 4^k.

iv. D(2) = 16

(b10 b11^2) 4^2 = 16^1

D(3) = 80

I (N_o + NO) 4^3 = 80

16b10 32b11 =16

b14 =64b10

192b11=80bn

Therefore $D(k) = (1/2 + (1/4)k)4^k = (2 + k)4$”

10.6. SOLUTION OF NONHOMOGENEOUS FINITE ORDER LINEAR RELATIONS

Our algorithm for nonhomogeneous relations will not be as complete as for the homogeneous case. This is due to the fad that different right-hand sides (f(k)'s) call for different rules in obtaining a particular solution.

Algorithm 10.6.1: Algorithm for Solving Nonhomogeneous Finite Order Linear Relations

To solve the recurrence relation 5(k) 5(k – 1) … C, 5(k – n) = f(k)

i. Write the associated homogeneous relation and find its general solution (Steps (a) through (c) of Algorithm 10.3.1). Call this the homogeneous solution, $5^{(h)}$ (k).

ii. Start to obtain a particular solution, $5^{0)}$) (k) of the recurrence relation by taking an educated guess at the form of a particular solution. For a large class of right-hand sides, this is not really a guess, because the particular solution is often the same type of function as f (k).

iii. Transform your second-step guess into a recurrence relation. If your guess was correct, you should be able to find the unknown coefficients. Return to Step 2 if your guess was incorrect, as the result of this substitution should indicate (Miller, 2011).

iv. The general solution to the recurrence relation is comprised of the homogeneous and particular solutions. If no conditions are listed, you are finished. If n initial conditions are supplied, the system will be solved to obtain a complete solution, which will convert to n linear equations in n unknowns.

Particular solutions for given right-hand sides.

Right hand side, f (k) Form of Particular Solution, $g^{(P)}$ (k)

Constant, q Constant, d

Linear function, go + q_1k Linear Function, $d_o + d_1k$

ra$^{t!}$' degree polynomial, flo + qtk “qin^{km}rnth degree polynomial, d_o d_1k + d,km

Example: Solution of a Nonhomogeneous First Order Recurrence Relation

Solve 5(k) + 55(k – 1) = 9, with S(0) = 6.

i. The associated homogeneous relation, 5(k) + 55(k – 1) = 0 has the characteristic equation a 5 = 0; therefore, a = –5. The homogeneous solution is $5(^h)$ (k) = $b(-5)^k$

ii. Since the right-hand side is a constant, we guess that the particular solution will be a constant, d.

iii. If we substitute. $5^{(\text{‘})}$ (k) = d into the recurrence relation, we get d + 5d = 9, or 6d = 9. Therefore, $S^{(P)}(k)$ = 1.5.

iv. The general solution of the recurrence relation is 5(k) = SO) (k) S(P) (k) = $-5)^k$ + 1.5 The initial condition will give us one equation to solve in order to determine b. 5(0) = 6 = b(–5)° ± 1.5 = 6 = b + 1.5 = 6 Therefore, b= 4.5 and S(k) = $4.5(-5)^k$ ± 1.5.

Example: Solution of a Nonhomogeneous Second Order Recurrence Relation

Consider T(k) – 77^1(k – 1) + 107^1(k – 2) = 6 + 8k with T(0) = 1 and T(1) = 2.

i. From Example 10.3.6, we know that $T^{(h)}(k) = b12^k\ b\ 5^k$. Caution: Don't apply the initial conditions to $7^1(^h)$ until you add T(‘)

ii. Since the right-hand side is a linear polynomial, T() is linear; that is, T(P)(k) = $d_0 + d_1$ k.

iii. Substitution into the recurrence relation yields: ($d_0 + d_1$ k) – 7 (do + di (k – 1)) 10 (d_0 + – 2)) = 6 + 8k

($4d_0 - 13d_1$) + ($4d_1$) k = 6 + 8k

Two polynomials are equal only if their coefficients are equal. Therefore,

Oda–13d1=61 Ido = 8

4di = 8 I t di = 2

iv. Use the general solution 21k) = $b12^k\ b25^k$ + 8 + 2k and the initial conditions to get a final solution:

f' TIM = 1 1 b2 + 8 =

t T(1) = 2 j 2b1 5b2 + 10 = 2

b2 = –7 $2b_1\ 5b_2$ = –8

b_1= –9

$b_2 = 2$

Therefore, 31k) = –9.2^k + 2 5^k + 8 + 2k.

Note: A multiplier is the simplest way to calculate the growth of a quantity, such as the balance of a savings account. In the instance of 8% growth, the multiplier is 1.08 since each original amount A has an addition of 0.08A, resulting in a new balance of A + 0.08A (1 + 0.08)A = 1.08A.

Another example is that if the interest rate were 3.5%, the multiplier would be 1.035. This assumes that the annual interest rate of 3.5%, sometimes known as simple interest, is applied at the end of the year. If the interest was applied monthly and we considered a simple example where each month was the same length, the multiplier after each month would be (1 (L*5) gt 1.00292. After a year, this multiplier would be applied 12 times, which is comparable to multiplying by 1.0029212 = 1.03557. Compound interest is to blame for the increase from 1.035 to 1.03557 (Miller, 2011).

Example: A Sort of Annuity

Suppose one open a savings account that can pay an annual interest rate of 8%. In addition, suppose you decide to deposit one dollar when you opened the account, and you now intend to double your deposit each year. Let B(k) be your balance after k years. B can be described by the relation B(k) = 1.08B(k – 1) + 2^k, with 5(0) = 1. If, instead of doubling the deposit each year, you deposited a constant amount, q, the 2^4 = term would be replaced with q. A sequence of regular deposits such as this is called a simple annuity.

Returning to the original situation,

i. $_B$o) $_{(k)}$ (1.08) k

ii. B$^{(1)}$ (k) should be of the form d2^k.d2^k = 1.08d2^{k-1} + 2^k = (2d)2^{k-1} = 1.08d2^{k-1} + 2 • 2^{k-1}

iii. 2d = 1.08d + 2.92d = 2

d = 2.174 to the nearest 1000th)

Therefore B$^{(P)}$ (k) = 2.174 • 2^k.d

B(0) = 1

b_1 + 2.174 = 1

b_i = –1.174

Therefore, B (k) = –1.174 • 1.08^k + 2.174 • 2^k.

Example: Matching Roots

Find the general solution to 5(k) – 35(k – 1) – 45'(k – 2) = 4^k.

i. The characteristic roots of the associated homogeneous relation are –1 and 4. Therefore, $5^{(h)}$ (k) = b1$(-1)^k$ b24^k

ii. A function of the form d4^k will not be a particular solution of the nonhomogeneous relation since it solves the associated homogeneous relation. When the right-hand side involves an exponential function with a base that equals a characteristic root, you should multiply your guess at a particular solution by k. Our guess at SIP) (k) would then be dk4^k (Miller, 2011).

iii. Substitute dk4^k into the recurrence relation for (k):

$dk4^k - 3d(k-1)4^{k-1} - 4d(k-2)4^{k-2} = 4^k$

$16dk4k^{-2} - 12d(k-1)4k^{-2} - 40\text{–} 2)4^{k-2} = 4^k$

Each term on the left-hand side has a factor of 4^{k-2}

$16dk - 12d(k-1) - 4d(k-2) = 4^2 20d = 16 = d = 0.8$

iv. Therefore, SO) (k) = 0.8k4^k

The general solution to the recurrence relation is

5(k) = b1$(-1)^k$ b24^k + 0.8k4^k

Observation: When the Base of Right-Hand Side is Equal to a Characteristic Root

If the right-hand side of a nonhomogeneous relation involves an exponential with base a, and a is also a characteristic root of multiplicity p, then multiply your guess at a particular solution by k P, where k is the index of the sequence.

Example: Examples of Matching Bases

i. if 5(k) – 95(k – 1) + 205(k – 2) = 2 5^k, the characteristic roots are 4 and 5. Since 5 matches the base of the right side, SIP) (k) will take the form dk5^k.

ii. If 5(n) – 65(n – 1) + 95'(n – 2) = 3'^{1}+^{1} the only characteristic root is 3, but it is a double root (multiplicity 2). Therefore, the form of the particular solution is dra^t 3n

iii. If Q(^{1}) – Q(j – 1) – 12Q(j – 2) = $(-3)^3$ + 6 the characteristic roots are –3 and 4. The form of the particular solution will be d_i j$(-3)^3$ d_2j • 4j.

iv. If 5(k) – 9S(k – 1) + 85(k – 2) = 9k + 1 = (9k + 1)1^k, the characteristic roots are 1 and 8. If the right-hand side is a polynomial, as it is in this case, then the exponential factor l^k can be introduced. The particular solution will take the form k (d_o + d_ik).

We finish this section with a comment on the case where the characteristic equation produces complex roots. Even if we limit the coefficients of our finite order linear relations to real numbers, or even integers, we can face characteristic equations with complex roots. We will merely point out that our methods are still valid with complicated characteristic roots, but the conventional approach for representing the solutions of these relations is different. We will just propose that an interested reader look to a more sophisticated discussion of recurrence relations (see also difference equations) because grasping these representations requires some knowledge in complex numbers (Miller, 2011).

Exercise 1:

5(k) –0.255(k– 1) = 0,5(0) = 6

Answer:

5(k) = 6(1/4)k

Exercise 2:

5(k) – 2S(k – 1) + S(k – 2) = 2, 5(0) = 25, 5(1) = **16**

Answer:

5(k) = k^2 – 10k + 25

Exercise 3:

5(k) – 5(k – 1) – 65(k – 2) = –30, 5(0) = 7, 5(1) = 6

Exercise 4:

S(k) – 5S(k – = 5^k, S(0) = 3

Answer:

S(k) = (3 ± k)5^k

Exercise 5:

5(k) – 55(k – 1) + 65(k – **2) =** 2, 5(0) = **1,**

Answer:

5(1) = 0

Exercise 6:

S(k) – 45(k – 1) + 45(k –2) = 3k + 2k, 5(0) = 1,5(1) = 1

Answer:

5(k) = (12 + 3k) + (k^2 + 7k – 22) 2^{k-1}

Exercise 6:

S(k)= r5(k – 1) +a, S(0) = 0, r, a > 0, T1

Find the answer.

Exercise 7:

5(k) – 45(k – 1) – 115(k – 2) 305(k – 3) = 0, 5(0) = 0,5(1) = –35, 5(2) = –85

Answer:

P(k) = 4(–3)k 2k – 5k+1

Find a closed form expression for the terms of the Fibonacci sequence.

The sequence C was defined by C_r = the number of strings of zeros and ones with length r having no consecutive zeros. Its recurrence relation is the same as that of the Fibonacci sequence. Determine a dosed form expression for C_r, r > 1 (Miller, 2011).

The characteristic equation is $a^2 - a - 1 = 0$, which has solution a = (I + $_v$i5)/2 and s=1 useful to point out that c + = 1 and a –/3 = vi5.

The general solution is F (k) = b_1 b_2s = I and $b_i b_2$i30)/2, It is Using the initial conditions, we obtain the system: $b_{2\,=}$ 8(c– = (5-v)/2/6 = ai(a – s) = (5+)/2 and b_2 an+1 sn+1

Therefore, the final solution is F(n) = a – (1 V)/2)$^{72+1}$ – N/75)/2r^{i}

b. C_r = F (8)

If 5(n) = i g(j), n > 1, then S can be described with the recurrence relation

5(n) = S(n – 1) + g(n).

For each of the following sequences that are defined using a summation, find a closed form expression:

- –S(n) = E 1j, n ^{1}
- Q(n) = $_1$ j^2 n > 1
- P(n) =E7–1 (1)^{i} n >0 d T(n) = E;L $_{1}$,> 1

Exercise 9:

Let D(n) be the number of ways that the set {1, 2. n}, n > 1, can be partitioned into two nonempty subsets.

- Find a recurrence relation for D. (Hint: It will be a first-order linear relation.)
- Solve the recurrence relation

Answer:

- For each two-block partition of {1, 2, n – 1}, there are two partitions we can create when we add n, but there is (Miller, 2011).
- One additional two-block partition to count for which one block is {n}. Therefore,
- D(n) = 2D(ri – 1) + 1 for n > 2 and D(1) = 0
- $D(n) = 2^{n-1} - 1$

Exercise 10:

An annuity is a series of payments made if a certain amount of money is placed at the end of each year for a number of years.

- It is necessary to find a closed form equation for the balance or value of an annuity with q payments of i dollars each at an interest rate of i. Keep in mind that a normal annuity's first payment is made after a year.
- How much would you need to put in an annuity with a 5.5% interest rate to make it worth $1 million after 18 years?
- A loan payment is a sort of annuity in which the principal amount of the loan has a negative initial value, and the annuity ends when the value is raised to zero. How much money could you borrow if you could afford to pay $5,000 each year for 25 years at an interest rate of 11%?

Exercise 11:

Suppose that C is a small positive number. Consider the recurrence relation:

$B(k) - 2B(k - 1) + (1 - C^2) - 2) = C^2$,

with initial conditions B(0) = 1 and B(1) = 1.

If C is small enough, we might consider approximating the relation by replacing $1 - C^2$ with 1 and C^2 with 0.

Solve the original relation and its approximation (Miller, 2011).

Let B_c, a be the solution of the approximation. Compare closed form expressions for B(k) and $R_a(k)$. Their forms are very different because the characteristic roots of the original relation were dose together and the approximation resulted in one double characteristic root. If characteristic roots of a relation are relatively far apart, this problem will not occur.

For example, compare the general solutions of S(k) 1.0015(k – 1) – 2.0040025(k – 2) = 0.0001 and Sa(k) Sa(k – 2,9,,(k – 2) = 0.

10.6.1. Solving Linear Homogeneous Recurrence Relations

Solving linear homogeneous recurrence relations of degree two – two distinct characteristic roots:

Definition: If $an = c1an–1 + c2an–2 + \cdots + ckan–k$, then

r

$k – c1r$

$k–1 – c^2r$

$k–2 – \cdots – ck–1r – ck = 0$ is the characteristic equation of an.

Ø **Theorem 1:** Let $c1$ and c^2 be real numbers. Suppose that r

$2 – c1r – c^2 = 0$ has two distinct

Roots $r1$ and r^2. Then the sequence $\{an\}$ is a solution to the recurrence relation

$an = c1an–1 + c^2an–2$ if and only if $an = \alpha1r1$

$n + \alpha^2r^2$

n

for $n = 0,1,2, \ldots$, where $\alpha1$ and $\alpha2$ are constants

Ø **Example 1:** What is the solution of the recurrence relation

$a^n = a^n–1 + 2a^n–2$ with $a0 = 2$ and $a1 = 7$?

- **Step 1:** Write the characteristic equation of a recurrence relation (CERR).

If $an = c1an–1 + c^2an–2$, then $r2 – c1r – c^2 = 0$ is the characteristic equation of an (Miller, 2011).

- **Step 2:** Solve CERR
- **Step 3:** Write solution in terms of αs.
- **Step 4:** Find $\alpha1$ and $\alpha2$ from the initial conditions.
- **Step 5:** Write the solution

Ø **Example 2:** Find an explicit formula for the Fibonacci numbers.

- · **Step 1:** Write the CERR.

If $an = c1an–1 + c^2an–2$, then r

$2 – c1r – c^2 = 0$ is the characteristic equation of an.

- **Step 2:** Solve CERR
- **Step 3:** Write solution in terms of αs.
- **Step 4:** Find $\alpha 1$ and $\alpha 2$ from the initial conditions.
- **Step 5:** Write the solution.

10.6.2. Solving Linear Homogeneous Recurrence Relations of Degree Two – One Characteristic Root of Multiplicity Two

Ø **Theorem 2:** Let $c1$ and $c2$ be real numbers with $c2 \neq 0$. Suppose that r

$2 - c1r - c^2 = 0$ has only one root $r0$. A sequence $\{an\}$ is a solution of the recurrence relation:

$an = c1an{-}1 +$

$c^2an{-}2$ if and only if $an = \alpha 1r0$

$n + \alpha 2nr0$

n

for $n = 0,1,2, \ldots,$ where $\alpha 1$ and $\alpha 2$ are constants.

Ø **Example 1:** What is the solution of the recurrence relation $an = 6an{-}1 - 9an{-}2$ with initial conditions $a0 = 1$ and $a1 = 6$?

- **Step 1:** Write the CERR.

If $an = c1an{-}1 + c^2an{-}2$, then r

$2 - c1r - c2 = 0$ is the characteristic equation of an.

- **Step 2:** Solve CERR
- **Step 3:** Write solution in terms of αs.
- **Step 4:** Find $\alpha 1$ and $\alpha 2$ from the initial conditions.
- **Step 5:** Write the solution

10.6.3. The General Case

Ø **Theorem 3:** Let $c1$, $c2$, ck be real numbers. Suppose that the characteristic equation (Miller, 2011)

r

$k - c1r$

$k{-}1 - c2r$

$k{-}2 - \cdots - ck{-}1r - ck = 0$

has k distinct roots $r1, r2, \ldots, rk$. Then a sequence $\{an$

$\}$ is a solution of the recurrence relation.

$an = c1an–1 + c2an–2 + \cdots ckan–k$

if and only if $an = \alpha1r1$

$n + \alpha2r2$

$n + \cdots + akrk$

n

for $n = 0, 1, 2, \ldots$, where $\alpha1, \alpha2, \ldots, \alpha k$ are constants.

Ø **Example 1:** Find the solution to the recurrence relation $an = 6an–1 – 11an–2 + 6an–3$ with $a0 = 2$, $a1 = 5$, and $a2 = 15$.

- **Step 1:** Write a CERR.

r

$k – c1r$

$k–1 – c2r$

$k–2 – \cdots – ck–1r – ck = 0$

- **Step 2:** Solve CERR.
- **Step 3:** Write solution in terms of αs.
- **Step 4:** Find αs from the initial conditions.
- **Step 5:** Write the solution

Ø **Theorem 4:** Let $c1, c2, \ldots, ck$be real numbers. Suppose that the characteristic equation:

r

$k – c1r$

$k–1 – c2r$

$k–2 – \cdots – ck–1r – ck = 0$ has t distinct roots $r1, r2, \ldots, rt$ with multiplicities $m1, m2, \ldots, mt$ respectively so that

$mi \geq 1$ for $i = 0, 1, 2, \ldots$, and $m1+m2 + \ldots + mt = k$. Then a sequence $\{an\}$ is a solution of the recurrence relation

$an = c1an–1 + c2an–2 + \cdots ckan–k$ if and only if

$an = (\alpha1,0 + \alpha1,1n + \cdots + \alpha1,m1–1n$

$m1–1)1$

$n + (\alpha2,0 + \alpha2,1n + \cdots + \alpha2,m2–1n$

$m2–1$

$)2$

n

$+ \cdots + (\alpha t,0 + \alpha t,1n + \cdots + \alpha t,mt–1n$

$mt–1$

$)rt$

n

for $n = 0, 1, 2, \ldots$, where $\alpha i,j$ are constants for $1 \leq i \leq t$ and $0 \leq j \leq mi–1$.

Example: Suppose that the roots of the characteristic equation of a linear homogeneous recurrence relation are 2, 2, 2, 5, 5, and 9. What is the form of the general solution?

Example: Find the solution to the recurrence relation $an = –3an–1 – 3an–2 – an–3$ with initial conditions $a0 = 1$, $a1 = –2$, and $a2 = –1$ (Miller, 2011).

- **Step 1:** Write a CERR.

r

$k – c1r$

$k–1 – c2r$

$k–2 – \cdots – ck–1r – ck = 0$

- **Step 2:** Solve CERR
- **Step 3:** Write solution in terms of αs.
- **Step 4:** Find αs from initial conditions.
- **Step 5:** Write the solution

10.6.4. Linear Nonhomogeneous Recurrence Relations with Constant Coefficients

Ø **Definition:** A linear nonhomogeneous recurrence relation with constant coefficients is a

recurrence relation of the form:

$an = c1an–1 + c2an–2 + \cdots + ckan–k + (n)$

where; $c1$, $c2$, …, ck are real numbers, and $f(n)$ is a function not identically zero depending only on n. The recurrence relation $an = c1an–1 + c2an–2 + \cdots + ckan–k$ is called the associated homogeneous recurrence relation.

Example: The following are linear nonhomogeneous recurrence relations with constant coefficients:

$a_n = a_{n-1} + 2^n$

$a_n = a_{n-1} + a_{n-2} + n^2 + n + 1$

$a_n = 3_{an-1} + n^{3n}$

$a_n = a_{n-1} + a_{n-2} + a_{n-3} + n$

Ø **Theorem 5:** If $\{a_n\ (^p)\}$ is a particular solution of the nonhomogeneous linear recurrence relation with constant coefficients

$an = c1an{-}1 + c2an{-}2 + \cdots + ckan{-}k + (n)$

Then every solution is of the form $\{an\ (p) + an\ (h)\}$, where $\{an(h)\}$ is a solution of the associated homogeneous recurrence relation $an = c1an{-}1 + c2a_{n-2} + \cdots + ckan{-}k$. Recurrence relations are important in mathematics because they give a powerful tool for describing and resolving a wide range of problems. Because they provide a rigorous strategy for defining the link between consecutive elements in a sequence or series, we can analyze and forecast their behavior. Many areas have applications for recurrence relations, including biology, economics, physics, computer science, and others (Miller, 2011).

By understanding the fundamental structures and properties of recurrence relations, mathematicians may build exact formulas, effective algorithms, and valuable predictions. In addition to improving our understanding of mathematical structures, the study of recurrence relations provides us with the tools we need to tackle difficult real-world problems by breaking them down into manageable, recursive steps.

Furthermore, by connecting several disciplines of mathematics, recurrence relations bridge the conceptual gap between discrete and continuous notions. They are widely used to build differential equations, combinatorial identities, and generating functions. This interplay between recurrence relations and different fields of mathematics emphasizes their relevance and versatility.

As we research and broaden our grasp of mathematics, recurrence relations will remain an important topic. Because of their beauty, versatility, and ability to discern subtle patterns, they are a crucial tool for tackling difficulties and broadening our understanding. The study of recurrence relations continues to affect how we approach and solve issues in mathematics, whether in pure mathematics or in practical applications (Miller, 2011).

Bibliography

1. Alibali, M. W., & Nathan, M. J., (2012). Embodiment in mathematics teaching and learning: Evidence from learners' and teachers' gestures. *Journal of the Learning Sciences*, *21*(2), 247–286.
2. Allenby, R. B., & Slomson, A., (2010). *How to Count: An Introduction to Combinatorics*. CRC Press.
3. Andrews, P. B., (2013). *An Introduction to Mathematical Logic and Type Theory: To Truth Through Proof* (Vol. 27). Springer Science & Business Media. *27*(4), 48–52.
4. Bacani, J. B., & Rabago, J. F. T., (2015). On linear recursive sequences with coefficients in arithmetic-geometric progressions. *arXiv preprint arXiv:1503.05301*.
5. Bell, M., & Schäfer, M., (2013). Semantic transparency: Challenges for distributional semantics. Available from: https://www.researchgate.net/figure/An-Eulerian-graph-the-weight-and-label-of-each-edge-is-given-on-the-figure_fig5_29636248 (accessed on 21 September 2023).
6. Ben-Ari, M., (2012). *Mathematical Logic for Computer Science*. Springer Science & Business Media. *15*(4), 77–84.
7. Bender, E. A., & Williamson, S. G., (2013). *Foundations of Combinatorics with Applications*. Courier Corporation. *11*(2), 104–109.
8. Benjamin, A. T., & Quinn, J. J., (2022). *Proofs That Really Count: The Art of Combinatorial Proof* (Vol. 27). American Mathematical Society. *14*(5), 108–111.

9. Benning, I., & Agyei, D., (2016). *Effect of Using Spreadsheet in Teaching Quadratic Functions on the Performance of Senior High School Students*, pp. 11–29. Available from: https://www.researchgate.net/figure/Graph-showing-the-translation-of-y-x-3-2_fig1_327424188 (accessed on 21 September 2023).
10. Berkolaiko, G., & Kuchment, P., (2013). *Introduction to Quantum Graphs* (No. 186). American Mathematical Soc. *20*(3), 81–88.
11. Burgos, M., Batanero, C., & Godino, J., (2022). *Algebraization Levels in the Study of Probability*. Available from: https://www.researchgate.net/figure/Probability-distribution-for-the-sum-of-two-dice-Algebraic-level-2_tbl1_357371988 (accessed 4 September 2023).
12. Cheng, P., (2020). Truth diagrams versus extant notations for propositional logic. *Journal of Logic, Language, and Information*. Available from: https://www.researchgate.net/figure/Truth-tables-1-4-relations-v-5-An-inference-over-formulas-Q-R-P_fig2_335775460 (accessed 4 September 2023)
13. Coecke, B., & Kissinger, A., (2018). Picturing quantum processes: A first course on quantum theory and diagrammatic reasoning. In: *Diagrammatic Representation and Inference: 10th International Conference, Diagrams 2018, Edinburgh, UK, Proceedings 10* (pp. 28–31). Springer International Publishing.
14. Cortés-Polo, D., Jimenez, L., Calle-Cancho, J., & González-Sánchez, J. L., (2019). *A Novel Methodology Based on Orthogonal Projections for a Mobile Network Data Set Analysis*. Available from: https://www.researchgate.net/figure/Two-dimensional-vector-space-representation-of-the-Euclidean-classifier-where-the_fig3_336865215 (accessed 4 September 2023)
15. Deisenroth, M. P., Faisal, A. A., & Ong, C. S., (2020). *Mathematics for Machine Learning*. Cambridge University Press.
16. Ďuriš, V., Bartková, R., & Tirpáková, A., (2021). *Several Limit Theorems on Fuzzy Quantum Space. Mathematics, 9*(4), 438. MDPI AG. Retrieved from: http://dx.doi.org/10.3390/math9040438 (accessed 4 September 2023).
17. Dwivedi, V. P., Joshi, C. K., Luu, A. T., Laurent, T., Bengio, Y., & Bresson, X., (2020). Benchmarking graph neural networks. *arXiv preprint arXiv:2003.00982*.

18. Dym, H., (2013). *Linear Algebra in Action* (Vol. 78). American Mathematical Soc. *12*(4), 80–84.
19. Efficiency of Random Swap Clustering, (2018). *Journal of Big Data, 5*, 10.1186/s40537-018-0122-y. Available from: https://www.researchgate.net/figure/Illustration-of-the-clustering-problem-using-a-pigeon-hole-principle-with-dataset-S2-The_fig1_324152730 (accessed on 21 September 2023).
20. Emeliyanenko, P., (2010). Modular resultant algorithm for graphics processors. In: *Algorithms and Architectures for Parallel Processing: 10th International Conference, ICA3PP 2010, Busan, Korea, Proceedings. Part I 10* (pp. 427–440). Springer Berlin Heidelberg.
21. Epp, S. S., (2010). *Discrete Mathematics with Applications*. Cengage Learning. *20*(5), 17–21.
22. Font, V., Godino, J. D., & Gallardo, J., (2013). The emergence of objects from mathematical practices. *Educational Studies in Mathematics*, *82*, 97–124.
23. Foulds, L. R., (2012). *Graph Theory Applications*. Springer Science & Business Media. *32*(8), 98–101.
24. Gill, P. E., Murray, W., & Wright, M. H., (2021). *Numerical Linear Algebra and Optimization*. Society for Industrial and Applied Mathematics. *22*(4), 118–124.
25. Grimmett, G., & Welsh, D. J., (2014). *Probability: An Introduction*. Oxford University Press.
26. Hager, W. W., (2021). *Applied Numerical Linear Algebra*. Society for industrial and applied mathematics. *14*(5), 108–111.
27. Haubold, H. J., Mathai, A. M., & Saxena, R. K., (2011). Mittag-Leffler functions and their applications. *Journal of Applied Mathematics*. *9*(2), 95–101.
28. Hunter, J. J., (2014). *Mathematical Techniques of Applied Probability: Discrete Time Models: Basic Theory* (Vol. 1). Academic Press.
29. Inglis, M., & Alcock, L., (2012). Expert and novice approaches to reading mathematical proofs. *Journal for Research in Mathematics Education*, *43*(4), 358–390.
30. Jukna, S., (2011). *Extremal Combinatorics: With Applications in Computer Science* (Vol. 571). Berlin: Springer.*14*(5), 108–111.
31. Kalimoldayev, M., Tynymbayev, S., Gnatyuk, S., Ibraimov, M., & Magzom, M., (2019). The device for multiplying polynomials modulo

an irreducible polynomial. *News of the National Academy of Sciences of the Republic of Kazakhstan, Series of Geology and Technical Sciences*, *2*(434), 199–205.

32. Kalnins, E. G., Kress, J. M., & Miller, W., (2011). A recurrence relation approach to higher order quantum superintegrability. *Symmetry, Integrability, and Geometry: Methods and Applications (SIGMA)*, *7*, 031.
33. Kochikov, I., Stepanova, A., & Kuramshina, G., (2022). Scaled in cartesian coordinates Ab Initio molecular force fields of DNA bases: Application to canonical pairs. *Molecules, 27*(2), 427. MDPI AG. Retrieved from: http://dx.doi.org/10.3390/molecules27020427 (accessed 4 September 2023)
34. Koekoek, R., Lesky, P. A., Swarttouw, R. F., Koekoek, R., Lesky, P. A., & Swarttouw, R. F., (2010). *Hypergeometric Orthogonal Polynomials* (pp. 183–253). Springer Berlin Heidelberg.
35. Kralik, P., & Ricchi, M., (2017). A basic guide to real time PCR in microbial diagnostics: Definitions, parameters, and everything. *Frontiers in Microbiology*, *8*, 108.
36. Kuwn, Y. C., Iqbal, Z., Nizami, A. R., Munir, M., Riaz, S., & Kang, S. M., (2018). Some Recurrence relations and Hilbert series of right-angled affine Artin monoid. *Journal of Function Spaces*. *31*(8), 98–112.
37. Lee, R. N., (2013). LiteRed 1.4: A powerful tool for the reduction of the multiloop integrals. *arXiv preprint arXiv:1310.1145*.
38. Lengyel, E., (2011). *Mathematics for 3D Game Programming and Computer Graphics*. Course Technology Press.
39. Levy, A., (2012). *Basic Set Theory*. Courier Corporation. *27*(4), 119–125.
40. Lockwood, E., (2013). A model of students' combinatorial thinking. *The Journal of Mathematical Behavior*, *32*(2), 251–265.
41. Luke, Y. L., (2014). *Mathematical Functions and Their Approximations*. Academic Press.
42. Ma, X., Dong, L., Wang, Y., Li, Y., & Sun, M., (2020). AIRC: Attentive implicit relation recommendation incorporating content information for bipartite graphs. *Mathematics, 8*(12), 2132. MDPI AG. Retrieved from: http://dx.doi.org/10.3390/math8122132 (accessed 4 September 2023).

43. Mane, S., Judge, L., & Schaumont, P., (2011). An integrated prime-field ECDLP hardware accelerator with high-performance modular arithmetic units. In: *2011 International Conference on Reconfigurable Computing and FPGAs* (pp. 198–203). IEEE.
44. Manin, Y. I., & Zilber, B., (2010). *A Course in Mathematical Logic for Mathematicians*. *4*(3), 13–17.
45. Mazur, D. R., (2022). *Combinatorics: A Guided Tour* (Vol. 55). American Mathematical Society. *7*(5), 120–128.
46. Monk, J. D., (2012). *Mathematical Logic* (Vol. 37). Springer Science & Business Media. *7*(8), 150–155.
47. Nundy, S., Kakar, A., & Bhutta, Z., (2021). *Understanding Medical Biostatistics*. Available from: https://www.researchgate.net/figure/A-normal-probability-distribution-curve-Figure-Downloaded-from-the-Internet_fig2_355487884 (accessed 4 September 2023).
48. Olver, F. W., (2010). *NIST Handbook of Mathematical Functions Hardback and CD-ROM*. Cambridge University Press.
49. Poole, D., (2014). *Linear Algebra: A Modern Introduction*. Cengage Learning. *3*(7), 78–81.
50. Quarteroni, A., Sacco, R., & Saleri, F., (2010). *Numerical Mathematics* (Vol. 37). Springer Science & Business Media. *6*(4), 88–91.
51. Romanovski, V. G., & Prešern, M., (2011). An approach to solving systems of polynomials via modular arithmetic with applications. *Journal of Computational and Applied Mathematics*, *236*(2), 196–208.
52. Schmidt, G., & Ströhlein, T., (2012). *Relations and Graphs: Discrete Mathematics for Computer Scientists*. Springer Science & Business Media. *10*(4), 27–34.
53. Shih, F. Y., (2017). *Image Processing and Mathematical Morphology: Fundamentals and Applications*. CRC Press.
54. Shiryaev, A. N., (2016). *Probability-1* (Vol. 95). Springer. *9*(7), 61–68.
55. Silva, A., Monticone, F., Castaldi, G., Galdi, V., Alù, A., & Engheta, N., (2014). Performing mathematical operations with metamaterials. *Science*, *343*(6167), 160–163.
56. Skudrzyk, E., (2012). *The Foundations of Acoustics: Basic Mathematics and Basic Acoustics*. Springer Science & Business Media. *17*(2), 22–25.

57. Speakman, M., & Courtney, S., (2021). *Exploring Matrix Multiplication with Manipulatives, 88*, 17–25. Available from: https://www.researchgate.net/figure/Matrix-multiplication-color-illustration_fig3_355007821 (accessed 4 September 2023).
58. Strang, G., (2022). *Introduction to Linear Algebra*. Wellesley-Cambridge Press.
59. Teypaz, N., & Rapine, C., (2008). *Graph Decomposition into Paths Under Length Constraints*. Available from: https://www.researchgate.net/figure/An-Eulerian-graph-The-weight-and-label-of-each-edge-is-given-on-the-figure_fig5_29636248 (accessed 4 September 2023).
60. Twomey, S., (2013). *Introduction to the Mathematics of Inversion in Remote Sensing and Indirect Measurements*. Elsevier. *15*(3), 102–109.
61. Wu, X., Yang, C. N., Zhuang, Y. T., & Hsu, S. C., (2018). Improving recovered image quality in secret image sharing by simple modular arithmetic. *Signal Processing: Image Communication*, *66*, 42–49.
62. Zimmermann, H. J., (2010). Fuzzy set theory. *Wiley Interdisciplinary Reviews: Computational Statistics*, *2*(3), 317–332.

INDEX

D

E